When the bough breaks...

When the bough breaks...

OUR CHILDREN, OUR ENVIRONMENT

Lloyd Timberlake • Laura Thomas

Earthscan Publications Ltd • London

First published in 1990 by
Earthscan Publications Ltd
3 Endsleigh Street, London WC1H 0DD

British Library Cataloguing in Publication Data
Timberlake, Lloyd
When the bough breaks: our children, our environment.
1. Environment. Conservation
I. Title II. Thomas, Laura
337.72

ISBN 1-85383-082-8

This publication accompanies the television documentary *When the bough breaks. . .* produced by Central-Observer in association with the Television Trust for the Environment, as part of Central ITV's Viewpoint '90 series.

Design by Mick Keates (071-381 6853)
Production by David Williams Associates (081-521 4130)
Typeset by DP Photosetting, Aylesbury, Bucks.
Printed and bound by Cox & Wyman

Earthscan Publications Ltd is an editorially independent and wholly owned subsidiary of the International Institute for Environment and Development (IIED).

CONTENTS

Authors' Note vii
Introduction. The Children Crisis 1
1. Children at the Centre 13
2. Women and Children Last 28
3. Polluting Posterity 62
4. Keeping Children Alive – from Calories to Cities 112
5. Medicine, Population and Power 155
6. Perilous Future – Global Warming and all that 182
7. Children and the Environment: Undervalued Resources 217
8. Remembering the Future 235
References 250

Hush-a-bye baby
on the tree top,
when the wind blows
the cradle will rock,
when the bough breaks
the cradle will fall,
and down will come baby and cradle and all.

Hush-a-bye baby
on the green land,
when the grass grows
the cradle will rock,
when the trees burn
the cradle will fall,
and down will come baby and cradle and all.

Hush-a-bye baby
under the sun,
when the clouds race
the cradle will rock,
when the air faints
the cradle will fall,
and down will come baby and cradle and all.

Hush-a-bye baby
on the sea shore,
when the waves run
the cradle will rock,
when the tide turns
the cradle will fall,
and down will come baby and cradle and all.

Hush-a-bye baby
on the cool stream,
when the weeds wave
the cradle will rock,
when the fish choke
the cradle will fall,
and down will come baby and cradle and all.

Hush-a-bye babies
sleeping tonight,
when the earth turns
your cradles will rock,
when the earth stops
your cradles will fall,
and down will come babies and cradles and all.

A poem especially written for
this book by Andrew Motion.

AUTHORS' NOTE

This book has had a complex birth. The United Nations Environment Programme (UNEP) and UNICEF have prepared for publication on 5 June 1990 – World Environment Day – a global report on *Children and the Environment*. Lloyd Timberlake worked with the Egyptian scientist Essam El-Hinnawi in the preparation of that document, which was also the report of the executive director of UNEP to governments for 1990.

Thus when Central Television, the *Observer* newspaper and Television Trust for the Environment joined forces to produce a television programme – the first of many planned environmental programmes – on children and the environment, based on the themes of the UN report, the idea was born for the publication of a popular book on the subject. Laura Thomas joined the effort as co-author.

While *When the bough breaks. . .* does not necessarily represent the views of UNICEF, UNEP or any other UN organization, and received no UN funding, we are grateful to both organizations for their encouragement, and for the data and information they so freely provided. (However, any errors in the book remain the sole responsibility of the authors.)

We are also grateful for the support we have received from Lawrence Moore and Robbie Stamp, producers of the television programme *When the bough breaks. . .* and the rest of the staff of Central Television, the *Observer*, Friends of the Earth, Save the Children, and all of the many other groups campaigning for a safer environment and for children's welfare, who have given us unselfish practical support and encouragement.

The North and the South

We often use the word "North" to refer to the richer, industrialized nations of Europe and North America (and Japan), and "South" to refer to the poorer, mainly tropical nations. The terms do not make much sense, as 85 per cent of the world's population live north of the equator, but they have become standard. We have also used the terms "Third World" and "developing nations" to refer to the Southern nations, even though all nations are "developing" in one sense or another.

Introduction
THE CHILDREN CRISIS

> 1. States Parties recognise that every child has the inherent right to life.
>
> 2. States Parties shall ensure to the maximum extent possible the survival and development of the child.
>
> *Article Six of the 1989 UN Convention on the Rights of the Child*

Every year, 14 million children under the age of five die in the developing world, not in a drought and famine year, but in an "ordinary" year.

Everyone talks of the "environment crisis" and the "population crisis"; most people have heard of the "debt crisis". But who ever speaks of the "children crisis"? It is as if these deaths have become an acceptable, muted, and natural background noise amid the louder workings of the world: buying, selling, trading, producing, governing, defending, and the general creation of wealth.

In one sense, most of these deaths can be linked to the environment. Four million die of diarrhoeal diseases, usually associated with bad water and contaminated food. Over five million die of diseases such as tetanus, whooping cough, measles and respiratory infections, all caused by microbes that are as "natural" and long-established in the human environment as the elephant or the oak. One million die of malaria, carried by another natural species, the mosquito. Others die of mixes of causes – often malnutrition and worms, which make young bodies vulnerable to opportunistic bacteria and viruses. Many of the most chronically malnourished children are in the care of parents who can no longer grow enough to eat in their over-used, bankrupt environments.

But in a more important sense, to describe these deaths as "environmental" is criminally misleading. The children are all the victims of political choices, and most of the remedies for their fatal conditions are absurdly cheap. If the world spent a fraction as much time and money on keeping children alive and managing the environment as it spends on arms, then the majority of children in jeopardy would survive, most forests could be saved and deserts could be kept in place.

The children's deaths and environmental destruction are also caused by the "crisis of under-development". This is not a crisis in the sense of an acute shock, but a chronic condition which has afflicted most nations for as long as they have been independent. It too has roots in political choices. The wealthy European and North American nations, have constructed an international economic system under which the rich countries get richer much faster than do most of the poor countries.

Farmers in the wealthy Northern nations are paid by their governments to produce crops which Europe and North America do not need and cannot sell. They are subsidized to use quantities of pesticides and fertilizers which pollute their own environments and drinking water supplies, and children are most vulnerable to that pollution. These crops are dumped on poorer nations in ways that do not feed the poor, but do give Third World governments little incentive to get their own agricultural houses in order.

Northern governments are slowly beginning to insist that the cost of the environmental damage caused by Northern manufacturing be included in the cost of their products. This is the "polluter pays principle". But the governments of the poorer Southern nations cannot charge Northern purchasers the cost of the environmental destruction caused by the production of lumber, cotton, sugar, peanuts, beef, tea, coffee and cocoa. So the poor South is subsidizing the rich North in selling these products, because the South assumes the costs of environmental destruction, costs which must be paid, initially by children alive today in those countries and then by future generations. Eventually, everyone everywhere will pay for the loss of tropical forests and topsoil.

The most obvious and most discussed sign of these politics of impoverishment is the debt crisis. During the four years of the

Marshall Plan following the Second World War, the United States sent two per cent of the wealth it produced every year to Europe to help it rebuild. For much of the 1980s, the indebted nations of Latin America and Africa have been sending twice that amount, in terms of a percentage of their own Gross National Products (GNPs), to the rich Northern nations every year, just to pay their debts.

Aid from most Northern nations has never amounted to more than a fraction of one per cent of GNP. Today the poor nations are shipping much more money to the rich nations than they get in return – through aid or in any other way.

The debt crisis means that Third World children are getting less to eat, less education and less health care. In many countries the progress in keeping children alive achieved by vaccination and better nutrition is being reversed. Children are paying their nations' debts with their futures. This is policy. Nations are paying these debts with their environmental resources, squandering topsoil, water and forests to raise hard cash.

So every day, members of parliaments and senates around the world, North and South, and members of government cabinets, North and South, are sitting down, receiving expert advice, debating, and deciding policies which guarantee that 14 million children will die and that nations will sink into environmental bankruptcy. The planet and those who live on it have come to a peculiar pass.

THE NEW ETHIC

There has been little recent change in humankind's ways of looking at children, but considerable change in its views of the environment.

For years we have kept the environment in a "policy ghetto", with some of those most concerned about the environment the

Overleaf Children are more vulnerable to the effects of pollution. Today's children and future generations are the guinea pigs in this generation's experiments with little tested industrial chemicals and pesticides. *Öko Freiburg*

most responsible. The very idea of "protecting the environment" suggests that human progress can somehow be kept separate from the human environment; the environment is a separate, sacred place; hands off. This has never been true in human history and never will be true.

All human progress or lack of it is based, ultimately, on use and misuse of the environment. It has traditionally provided clean water, clean air, and protection against the sun's most harmful rays. It has provided a predictable climate for agriculture, adequate soil and water, and other natural resources. The ways in which human societies have chosen to develop – policies again – mean that the continued supply of these services can no longer be guaranteed. And there are three more people, all of whom need them, added each second to the world's population.

It is not just that "the environment" has become more important as a political issue. The change has been much more profound. We have come to realize, finally, that we cannot progress in ways which destroy the environment, as the environment is the basis of all progress. So the very nature and meaning of progress have become important political issues. Concern for the needs of future generations is beginning to guide us in the use of resources and the use of the environment. We must, we are told, live today not only so that we meet our own needs, but so that we leave adequate resources for future generations. Future technology, an unknown in the equation, may help future generations use resources more efficiently, but bigger populations will almost certainly see to it that our progeny require more resources.

The fatal flaw in this reasoning is that policymakers and people in general have never been required to concern themselves logically, mathematically, economically and politically with the needs of future generations. Given the apparently limitless

Opposite A girl looks after her younger sister in an Indian village. Some 85 per cent of the world's children live in the Third World, where environmental destruction is most threatening.
Mark Edwards/Still Pictures

bounty of the environment, any wealth creation at a given time was seen to be wealth creation for the future, which would also have infinite resources.

So, from almost a standing start, political leaders are trying to find ways of investing taxpayers' money in the needs of next century's taxpayers – while at the same time, trying to win the next election. Jurists are trying to establish legal principles which will guarantee a sort of equity between this generation and future generations. Economists, who have always discounted future against present benefits, are trying to find ways of spreading their spreadsheets wide enough to cover future needs.

If our main guide in the rational management of the environment is to be the needs of future generations, then – based on present practices – we have very little chance of ever managing our environment wisely. To reach this conclusion, one need not peer far into the future. Look instead at how we meet the needs of the very next "future generation", the large flesh-and-blood generation which lives in our midst today: the children of the world. The 14 million deaths each year, the systematic ways in which children's lives are destroyed as governments strive to pay and to collect debts, the low priority given to the needs of children in many wealthy countries, do not inspire confidence that we can meet any future needs.

Children are of course much better off in Europe and North America than in the Third World because their first line of defence and caring, their parents, are much better off. They have more political and economic power than most Third World parents. There are safety nets in the forms of state pensions, sick pay, unemployment benefits, child allowances, compulsory free education, and national health services.

Yet, though it is a matter of degree, Northern governments tend to discount the future of children. Over the past decade, the number of children in the United States living below the poverty line has increased by three million. The United States, the world's wealthiest nation, has a higher infant death rate than most European nations.

In Britain, the number of homeless families has doubled over the past decade. It has taken European governments far too long to begin to get lead out of petrol, and thus out of the blood and brains of children, who are most severely hurt by it. Today, hundreds of thousands of British households are still drinking water containing dangerous levels of lead. Britain allows fruit to be sprayed with chemicals withdrawn in other nations because they are suspected of causing cancer. Children eat more fruit than adults, and because of the ways their young bodies work they are more vulnerable to all pollutants, cancer-causing and otherwise. The breast milk of British mothers contains 100 times more dioxin than the government's recommended safety guideline.

Children are also exploited in the richest nations. The world's most deadly drug, tobacco, is routinely sold to them. Advertisements and commercial pressure aimed directly at them through television, newspapers and magazines teach them to want more of everything, despite the fact that the rich nations are already consuming far more than their fair share of the world's resources, including its natural resources.

Children in Eastern Europe suffer much more acutely from decades of unchecked pollution and environmental mismanagement, as central planning failed to plan for them a healthy present or a liveable future. It was a combined concern for children and for the environment, as well as for human rights, which led to the recent upheavals across Central Europe and the Soviet Union.

Policymakers cannot make rational policies for future generations unless they can make rational policies for today's children. So all the threats to children, and all the abuses they suffer, provide a measure of this generation's ability to provide for the needs, including the environmental needs, of future people. And this generation is not measuring up. This book, therefore, covers not only the ways in which children suffer from environmental mismanagement; it touches upon other forms of the mismanagement of children's present and future hopes.

THE NEXT GENERATION

There is no agreed international definition of a "child". The 1989 UN Convention on the Rights of the Child sets the upper age at 18, unless national laws set a different age for "majority", although children of 15 and above are declared old enough to fight in wars. UNICEF tends to define those under 16 as children.

Today, about one third of the planet's population is under the age of 16. The vast majority of these children, over 85 per cent, live in the developing world. Asia alone – with over one billion children – has as many children as China has people. Asia also has half of the world's children who live in absolute poverty.

Europe (excluding the Soviet Union) has 49 million more people than Latin America; but Latin America, because of its rapid population growth rates, has 66 million more children than Europe.

So sheer numbers – and population growth rates are also associated with govenments' policies – make it harder for the Third World to care for its children. And nine-tenths of all future population growth is expected to take place in the developing nations.

The regions where most of the world's children live suffer rapid soil degradation, forest loss and water scarcities. These regions also contain the greatest numbers of wild species of plants and animals, the "genetic diversity" which is being rapidly wasted, depriving future generations of crops, medicines and chemicals. Natural resources will stand little chance against the needs of millions, unless these needs can be managed.

Global warming and ozone depletion will hit these children-filled regions the hardest. The "greenhouse" effects are expected to be felt first in terms of a growing number of disasters such as floods, droughts and cyclones, and children always suffer disproportionately in disasters of any kind. More chronic effects will be an increased spread of deserts and deforestation, shrinking harvests and higher levels of malnutrition. The pattern of diseases, many of them most dangerous to children, is expected to change, further burdening already over-stretched healthcare systems. There is even the fear that increased radiation due to loss of the ozone layer will weaken human immune systems.

Amid all the gloom is evidence of real progress. Vaccinations and a simple sugar-and-salt solution to keep children alive during bouts of diarrhoea are estimated to save three million children each year. The technology, even delivering the technology, is cheap. But progress to date is fragile. It relies too much on the supply of technology from outside the societies concerned. The debt crisis is making it hard to maintain this progress, and if the worst fears of global warming and ozone depletion are realized, it could so easily be reversed.

THE NEW COALITION

Governments, even the best, have trouble investing in children or the environment because they live only from election to election. Those that do not face elections – the majority of governments – must answer to the demands of a small minority of elites: the military, the civil servants and business interests. Money invested in children and the environment is money invested today which will realize only vague and unreckonable future benefits. More important, many such benefits will go to voters of the future rather than those of today.

The children of government leaders do not suffer from lack of government spending on health care and education. Nor do the children of government leaders suffer the worst effects of environmental destruction. Governments would rather invest against more concrete, traditional threats, such as military attack, despite the diminishing probability of those threats materializing.

So motivation for change must lie outside governments, but it must eventually galvanize them into action. Only they, co-operating with one another as never before in history, can protect humanity against the planetary threats of global warming, ozone depletion and the loss of forests, species and productive lands.

The new concern for the environment, and thus for the future and for children, promises to create a new coalition of concerned citizens. At last, the many single-issue pressure groups can be united under one banner. The environmentalists, those concerned with development for the poorest, women's rights groups, peace campaigners, tribal rights groups, children's rights groups, those concerned with international co-operation: all have a stake in future generations. There are already signs that such coalitions are forming.

1
CHILDREN AT THE CENTRE

> We borrow environmental capital from future generations with no intention or prospect of repaying. They may damn us for our spendthrift ways, but they can never collect our debt to them.
>
> *World Commission on Environment and Development*

Children are beginning to muscle their way into the centre of the debate on how to manage the environment. They and their well-being increasingly provide a report card on how humankind is progressing in its attempts to preserve the planet for all future generations.

Few have noticed this, largely because the controversy about the environment in particular and all human progress in general has changed so radically in only a few years. The thinking among scientists and government officials is very different in 1990 from what it was in 1985. Problems like global warming, the hole in the ozone layer, acid rain, and the loss of rainforests and their plants and animals have all meant that the environment has stopped being the concern only of the "environmentalists" and has become a real concern, finally, of politicians.

Any attempt to make this concern rational must focus on the welfare of future generations, and thus must concentrate on children. The 1990s will see a wave of reports, books, articles, television programmes and conferences on children and the environment.

To understand this shift in perception, it is necessary to look back to the first wave of environmental anguish and action in the late 1960s and early 1970s.

That wave was set rolling mainly by rich people in rich countries. The radical youth of the sixties used environmental destruction as one weapon with which to attack "the Establishment". But older people with the leisure time to ramble in the countryside, to watch birds, to hunt ducks and to go boating also noticed that all the industrial progress which had paid for their leisure was destroying the countryside in which they liked to spend it.

Most of the early worry was local: polluted rivers and lakes, dirty air, urban blight, birds being poisoned by pesticides. But as the evidence mounted, concern became national and then international: global warming and ozone depletion (even back then), acid rain, vanishing species and dirty oceans. Many nations passed Clean Air and Clean Water Acts; ministries of the environment sprang up around the world. In 1972, the Earth's governments gathered in Stockholm for one of the first big global conferences on a single issue: the UN Conference on the Human Environment. It was also in 1972 that the Club of Rome published its grim *Limits to Growth* report, which used computer projections to show that human progress would soon be limited by pollution and a lack of natural resources.

The poorer countries came along to Stockholm, but were largely unconvinced. They felt that they had too much unspoiled environment, and not nearly enough economic progress. After all, Britain had cleared nearly all its forests and the United States had ploughed all its virgin prairie to make way for farming. Europe and North America had polluted their water and air to manufacture things their people wanted. Were others to be denied similar "progress"? The developing countries were worried that aid-giving nations would actually stop giving aid for power stations and ranches in the name of the environment. "Give us some of your pollution, as long as we can have the industry that goes with it", a Brazilian delegate to Stockholm was quoted as saying – not from the podium, but in the corridors.

In the North, voter pressure, media attention and campaigns by green groups encouraged new laws which led to improvements. The air became more breathable in most cities; lakes and rivers were cleaned up. Some nations, such as the United States, began to take lead out of petrol.

But during the course of the 1970s, the economic environment changed more radically than the natural environment. Economies slumped and oil price rises slowed energy use. It began to seem to governments that economic problems were much more limiting to growth than environmental dead ends. Besides, the Club of Rome report was being widely discredited for not taking account of technological progress.

This change in priorities was justified by many of the new market-oriented Northern leaders of the early 1980s, who felt that the environmental battle had been fought and won. Ronald Reagan worked systematically to dismantle much of the regulating clout of his government's Environmental Protection Agency, one of the most powerful agencies ever established in that country, by appointing anti-environmentalists to run it. Margaret Thatcher became prime minister of Britain in 1979, and did not mention the environment as an issue worthy of concern until late 1988.

By 1982, speeches celebrating the tenth anniversary of the Stockholm conference praised the industrialized nations' progress in cleaning up. There emerged the vague notion among ordinary citizens that the environment was somehow under control. This was accompanied by the equally vague and much more dangerous notion that the environment could always be cleaned up, whenever governments were ready, once the time to clean it up had rolled around.

But at precisely the same time, two new types of concerns were emerging which painted a darker picture. First, those same tenth anniversary speeches by the experts said that while the North had achieved some progress, the focus of environmental damage had shifted southwards into the developing world. Tropical forests were vanishing, particularly in Latin America

Overleaf Children along the Yamuna River near Delhi. Three pumps are meant to provide drinking water for the neighbourhood. One has gone dry; the other two have caused cholera epidemics over the past two summers.
Meera Dewan/Central Independent Television PLC

and Asia, to make way for unproductive and unprofitable farms and ranches. Good land was being turned into unproductive desert through misuse and over-use, particularly in Africa. The droughts and famines of 1984–86 across Africa showed that this was a human, more than an environmental, tragedy. Gradually leaders both inside and outside the affected nations began to see this destruction as dampening hopes for economic progress in the mainly agricultural South. They also saw that genetic diversity, with its promise of new medicines and crop species, was being lost with the forests, and this loss affected the whole world.

Second, the rich nations began to grow more and more worried about international threats like acid rain, global warming and ozone loss. The scientific data on these threats had been poor at the time of Stockholm. While there had been some progress on national pollution by 1982, there had been no real progress on the global issues. And governments were beginning to see that there could be no progress without international co-operation.

THE WORLD COMMISSION

With these developments in mind, in late 1983 the United Nations established an independent commission. Its brief was ambitious. It was supposed, among other things, to "propose long-term environmental strategies for achieving sustainable development by the year 2000 and beyond" and to "recommend ways concern for the environment could be translated into greater co-operation among developing countries and between countries at different stages of social and economic development" (that is, between the Rich and the Poor).

The UN appointed a unique politician to head this new "World Commission on Environment and Development". Gro Harlem Brundtland had been environment minister of Norway, and from that post she had gone on to become prime minister. She likes to claim that she is the only politician ever to rise from the traditionally thankless and dead-end job of the environment portfolio to lead a nation, and that this gives her an insight most political leaders lack. Even more important, for the purposes of the new commission, she had lost the prime minister's job, and had time to devote to the commission.

In 1986, in the middle of the commission's work, she became prime minister again, which forced greater respect from other countries for her and her commission as they travelled the globe studying the situation. It is also worth noting, given the effect that her commission's report has had on the issue of children's welfare, that Brundtland was trained as a paediatrician and a professional in children's public health.

Unlike previous commissions, such as Willy Brandt's (former Chancellor of West Germany) on North–South issues and the late Prime Minister of Sweden Olaf Palme's on security and disarmament, this group held public hearings on five continents, listening to farmers and rubber tappers and leaders of citizen's groups, as well as to scientists. The three-year education for 22 commissioners cost about $6 million, and it changed the preconceptions of many of them. Sir Shridath Ramphal, Secretary General of the Commonwealth of Nations, knew all about the effects of poverty in poor nations; he emerged making speeches about environmental destruction adding to that poverty. William Ruckelshaus, a lawyer who had twice headed the US Environmental Protection Agency, knew all about pollution, but emerged trying to convince his fellow Americans that global poverty was their problem as well.

The 22 commissioners were a good mix of West and East, North and South, but not of male and female – there were only three women, including Brundtland. As a group, they were elderly, and they were all from the elite of their nations: an Italian senator, the Zimbabwe finance minister, the Indian president of the International Court of Justice, the Indonesian minister of population and environment, the former Nigerian minister of agriculture, and so on.

They published their final report, *Our Common Future*, in early 1987, and presented it to the United Nations General Assembly in October of that year – on the same Monday morning that the New York stock market crashed, a coincidence which meant that the presentation and all the fine speeches by heads of state on the day got almost no coverage by the media.[1]

At the same time, the UN Environment Programme (UNEP) presented to the General Assembly a very similar report entitled *Environmental Perspective to the Year 2000 and Beyond*, containing parallel conclusions. The World Commission, though established by the United Nations, was not a UN commission; it was "independent" and its views were supposed to represent the independent conclusions of the commissioners, not necessarily of their governments. By presenting its own report, UNEP produced an official UN document to which there had to be an official response. There were resolutions passed in late 1987 hailing both reports as the way forward for the UN system in particular and for all governments in general.

CONCLUSION: WE CAN'T GO ON THIS WAY

Our Common Future is written in the cautious language one might expect of a unanimous report by 22 senior bureaucrats. It is this caution, and the fact that it was unanimous, that gives it such power. For hidden in the careful language are startling conclusions – startling not because they were new, but because of the nature of the people who were saying them.

The most startling of these was that the ways in which we are doing many basic things – producing energy, growing food, manufacturing goods, protecting wild species – offer little hope for the future. They are *unsustainable*; we cannot keep doing things the way we are and hope to hand over a safe and sound world to our descendants. Environmentalists have been saying this for years; in *Our Common Future*, people with real power and real responsibilities began to say it too, in unison, and for the first time.

Energy offers one of many examples. In the early 1980s, the world population of almost five billion was annually consuming energy equivalent to the burning of 10 billion tons of coal. Most of it came from fossil fuels such as coal, oil and gas, and, in the South, wood and charcoal. This level of consumption is slowly warming the planet, producing acid rain in Europe and North America, and accelerating deforestation in the tropics.

If present energy use per person were to remain constant, then a world of 8.2 billion people in the year 2025 would need the equivalent of 14 billion tons of coal, over four billion in the South and over nine billion in the North.

But this projection is based on present consumption rates, heavily biased towards the North. What about "development", giving the South, where three-quarters of the world's people live, a fairer share? If the people of the South began to consume energy at the same rate, per person, as the people of the North, then by 2025 the world would need energy equivalent to the burning of 55 billion tons of coal. Clearly neither the atmosphere nor the people could stand the pollution produced by such energy consumption. This sort of back-of-envelope calculation leads to a choice between two conclusions. Either the "developing world" can never develop in the way in which it has always expected to, that is, catching up with the North in terms of cars, fridges, lights, factories and power plants; or new forms of energy production are needed.

The commission conducted similar exercises for industry, agriculture, forest use, fishing and other areas of production. So it is small wonder that the major theme running through its 400-page report was "sustainability" or "sustainable development". How can we begin to make genuine progress, to realize human needs and human wants, in ways that are sustainable and that offer the planet a hopeful future? The commissioners were never happy with the term "sustainable development", because it seemed to imply a prescription for the "developing" world, rather than for the whole world. But they comforted themselves with the notion that all nations, even the richest, are still developing in one direction or another; so sustainable development should really be thought of as sustainable human progress.

FOCUS ON FUTURE PEOPLE

Taking a general, global view and not wanting to offer a "blueprint" for individual nations, the commissioners defined sustainable development in only the most general, global terms: "it meets the needs of the present without compromising the ability of future generations to meet their own needs".

"Sustainable development" is hardly a riveting battle cry. It lacks the zest of such slogans as "Liberty!" or "Equality!". It is also vague. What does it really mean in terms of day-to-day activities? But it has in a few years had the effect of taking concern for the environment out of the Green ghetto, and has brought economists, political scientists, sociologists and industrialists into the debate. Along with the thinking behind it, the concept of sustainable development has convinced many previously doubtful people that concern for human progress and concern for the environment cannot be separated.

"In an important sense, there has turned out to be more to the notion of sustainable development than even the wise members of the World Commission intended", wrote Professor William Clark of Harvard University. "Individuals, organizations and entire nations have taken the concept as a point of departure for rethinking their interactions with the global environment".[2]

But the Commission's definition, based on the needs of future generations, presents a huge and as yet little discussed challenge to humanity. It also puts children at the centre of the whole environment/development debate.

We are asked to conduct our present affairs with one eye on the needs of future generations. But we have never had to do that before. Despite setbacks such as the Black Death and world wars, there has long existed in the collective human consciousness the idea of a benign *progress*. Developments in science, technology and medicine were seen to be producing a safer, more comfortable world. This optimism was particularly strong during the rebuilding period after the Second World War. Recessions were seen as temporary dips on an upward curve. Concern for future generations has been limited to educating and providing for our children so that they would be well placed to use the fruits of progress.

It is only relatively recently, perhaps first seen in fears of global destruction by nuclear weapons, that this faith in progress on the part of ordinary people has begun to flag. The growing awareness of both increasing world poverty and environmental destruction has played a part. The mavericks and poets noticed first. The social critic Lewis Mumford wrote that, "Today, the notion of progress in a single line without goal or limit seems perhaps the most parochial notion of a very parochial century". The Argentinian writer Jorge Luis Borges maintained, "We have stopped believing in progress. What progress that is!"

Present threats to global ecosystems have raised the fear that not only are we not progressing towards a more comfortable world, but we are creating a dangerous and unpredictable world for our progeny. Humanity has never faced this prospect before. According to *Our Common Future*:

> When the century began, neither human numbers nor technology had the power to radically alter planetary systems. As the century closes, not only do vastly increased human numbers and their activities have that power, but major, unintended changes are occurring in the atmosphere, in soils, in waters, among plants and animals, and in the relationships among all of these.

So we are told that we are destroying the world as we and our forebears have known it, a world for which our activities seemed so well designed, and creating a perilous and unpredictable world in which our present methods of producing wealth are unlikely to prosper. And we are told that the only way out, the only rational guideline we have to follow, is to commit ourselves to the needs of future generations. Without that concern for future people, we have no reason to stop living as we do.

But we have no practice in such concern; we have no institutions that represent future generations; they have no voice in our deliberations. The commissioners themselves, all coming from the world's premier political and economic institutions, admitted that the rapid rate of change "is frustrating the attempts of political and economic institutions, which evolved in a

Overleaf What future for these children in a shantytown in Lima, Peru? Some 90 per cent of future population growth will be in the Third World, and the Third World is rapidly moving into cities.
Mark Edwards/Still Pictures

different, more fragmented world, to adapt and cope". This admission also leads to a choice between two difficult conclusions: either humankind's political and economic institutions must adapt themselves and their decisions to the new reality, or humankind will suffer disastrously.

"Future generations" is a vague and abstract notion. Who has any vision of the world in which our great-great-grandchildren will live, or how they will live in it? But we need not look that far ahead.

Today's toddlers will only be middle-aged by the year 2030. That is the year by which all the various greenhouse gases in the atmosphere are expected to have an effect equal to a doubling of the carbon dioxide concentration present before the Industrial Revolution. That doubling is expected to increase global temperatures, raise sea-levels and result in more droughts, floods, storms, cyclones and general social disruption.

The first effects of pollution-driven global warming may well be upon us today. The six warmest years since record-keeping began occurred in the 1980s, a decade which also saw droughts in Africa, the United States and elsewhere, fierce hurricanes, peculiar monsoons in Asia, and destructive winter storms in southern England in 1987 and 1990. In no single case can the hand of the greenhouse effect be proved to have been at work. But James Hansen, at the Goddard Institute for Space Studies in New York, was willing to testify before a US Senate committee as early as 1988 that there was only a one per cent chance that temperature increases seen in the past few years were accidental.[3]

The notion of sustainable development brings children into the centre of the environment/development debate. The needs of children alive today for a safe environment are not being met. They may be the first real "greenhouse generation". Their own children, our grandchildren, will be particularly at risk. The immature of the species *Homo sapiens*, like the immature of any species, are the most vulnerable to sudden shocks and changes. They suffer most and first from environmental degradation, and their suffering will deepen as these changes accelerate.

The World Commission wrote:

We borrow environmental capital from future generations with no intention or prospect of repaying. They may damn us for our spendthrift ways, but they can never collect our debt to them. We act as we do because we can get away with it: future generations do not vote; they have no political or financial power; they cannot challenge our decisions. But the results of the present profligacy are rapidly closing the options for future generations. Most of today's decision makers will be dead before the planet feels the heavier effects of acid precipitation, global warming, ozone depletion, or widespread desertification and species loss. Most of the young voters of today will still be alive.

However, it is not just children's environmental welfare but children's overall welfare which becomes the standard by which we must judge our ability to achieve sustainable development.

In children alive today, we have among us the first representatives of the future generations which underpin the notion of sustainable development. Any evidence that we undervalue them is evidence that we undervalue our future as a species on earth. Thus evidence that we cannot care for them is evidence that we cannot care for our environment.

Such evidence abounds. In times of economic hardship, we cut back on education and health care for the poorest children. We exploit them, commercially and sexually. Some nations use them to fight their wars. We allow them to roam homeless in our cities. The most damning evidence against us is the fact of the 14 million children under the age of five who die every year in the Third World – deaths which could relatively easily and cheaply be prevented.

If we cannot build the needs of millions of children into our political, economic and social systems, we have little chance of meeting our own and our descendants' needs for a sustaining and sustainable environment.

2
WOMAN AND CHILDREN LAST

> They have in their hands youth and vigour and consequently the wind and favour of the world behind them.
>
> *Montaigne*

This thing the world calls the "debt crisis" almost killed 11-year-old Pablo Mamani and his family. Today he is out of school, shining shoes, with nothing much to look forward to. But he is alive.

Pablo's father, Eusavious, is a mountain Indian who worked in the early 1980s in the "Siglo XX" (Twentieth Century) tin mine 4,000 metres up in the Bolivian Andes north of La Paz. Like all mines in Bolivia, it was a government mine. It offered the miners one of the best hospitals in the country, social security, a store that sold food at subsidized rates and, most important from Pablo's point of view, an excellent secondary school. If he had completed that school, he would not only have had a diploma, but would have been taught a skilled craft such as plumbing or carpentry. He never started at the school.

In 1985, with annual inflation running at 20,561 per cent, Dr Victor Paz Estenssoro was elected president. He had already been president once, between 1952 and 1956, and had started a

Opposite A Bolivian miner from the high Andes, thrown out of work so their country can pay its debts, trying to clear Amazonian rainforests. *Mark Edwards/Still Pictures*

"workers' revolution", including nationalization of the mines. When he came back in 1985, he began one of the most drastic economic "adjustment" programmes ever attempted in Latin America. About half of all government workers in La Paz were fired. Unproductive mines, including Siglo XX, were closed. Eusavious and 4,500 colleagues were thrown out of work. The government gave them land to farm in Magdalena in the Bolivian Amazon forest to the east.

Eusavious took his wife, Flora, and Pablo into the jungle to clear a farm. Their people are descended from the Incas; they are used to temperatures ranging from –8°C to 14°C, and dry, thin air. In their new home, the temperatures ranged from 14°C to 45°C, with killing humidity. No doctors were sent in with the miners; they got no vaccinations or health care; they caught yellow fever, malaria, and other diseases against which their bodies had no natural defences. They got no training in jungle farming. They were not even meant to grow cash crops; the government simply wanted them to colonize an underpopulated region, to spread the Bolivian nation into the jungle.

Many died, quickly. Those who survived found that the soil lost its fertility after only one year. They had either to clear more forest or to leave. Almost all left, moving into the shanty towns of La Paz, where they sell such things as candy and pencils on the streets. The little clearings in Magdalena are deserted wastelands.

Eusavious was more enterprising. He has returned to the mountain mining town of Llallagua, where he lived before. It is a virtual ghost town, with perhaps 20 per cent of its former population. All the government buildings – almost all buildings – are shut. Eusavious now works in a "co-operative" mine, which means that he and his colleagues go underground every day, and, using whatever tools they can salvage, scrape out tin. There is no support, no medical care, no school, no cheap food, no sick pay or social security.

Flora washes glasses in a café. Pablo shines shoes, though there are not many people around who want their shoes shined. Pablo does not understand the debt crisis, though he has lived on its cutting edge for five years. His family, like millions thoughout Latin America and Africa, lost their livelihoods and tried to survive by living off environmental resources. The Mamani family failed, but are getting along. Pablo's future has been foreclosed to pay his nation's debts.

"MUST WE STARVE OUR CHILDREN . . . ?"

In 1986, former Tanzanian President Julius Nyerere asked one of the most critical questions of the decade: "Must we starve our children to pay our debts?" It is not clear that the world gave this very serious question any very serious thought, but the answer proved to be an unequivocal "Yes".

But why was infant malnutrition rising in much of the Third World at the *end* of the 1980s? Why was there less money for children's feeding, education and health care? Why were rising numbers of children left to wander the streets?

The popular image of the 1980s among Europeans and North Americans is of a decade of natural disasters in the developing world, primarily the droughts and famines in Africa in the mid-1980s, but also floods in Bangladesh and Mozambique, hurricanes in the Caribbean and an earthquake in Mexico. People in the North put on their running shoes and dug into their pockets and gave unprecedented amounts of money to Band Aid, Live Aid, Comic Relief and to the older development charities such as Oxfam and Save the Children. The budgets and staff of the larger agencies, and the number of projects they were running in the Third World, increased several-fold.

Surely this improved things? And if it did not, was it because the money was spent unwisely? Even Britain's sophisticated *Financial Times*, examining the record of Band Aid after it had wound up operations in 1989, seemed to suggest that there was some connection between Band Aid's work and the fact that Ethiopia was still in trouble: "In the end, any assessment of Band Aid's operation is likely to be inconclusive. On the relief front, Band Aid saved lives as it intended. But in Ethiopia despite all aid efforts and good rains in 1988, the country is still in a crisis".[1]

Most of Ethiopia's troubles stem from a corrupt and muddled government and eternal civil wars, but the country also owed debts totalling almost $2.5 billion in 1987; small by global standards, but equal to half the nation's annual Gross National Product (GNP – the value of all goods and services produced). That year it was expected to pay $51 million in interest on its debts, more than one quarter of the value of the goods and services it exported.

Sub-Saharan Africa as a whole is in worse shape. In 1988 it owed $134 billion, about equal to the region's Gross National Product. Its annual obligations to "service" those debts equalled almost half of what it was earning from exports. Since interest payments and debt repayments are in "hard" currencies such as dollars and sterling, and since these nations earn hard currencies mainly by exports, servicing the debts means there is little left for key imports. No more than a dozen sub-Saharan nations have regularly serviced their debts since 1980.[2] Africa's debts are crushing to African governments, but small compared to those of many Latin American nations.

The tragic bottom line is that during the 1980s, the poorest nations were shipping money, the produce of their over-used environments and their childrens' hopes for the future, to the richest nations. They were sending north much more than they were getting in return.

At the beginning of the decade, the developing nations were receiving a net total from the North of $40 billion. Today, taking into account loans, aid, repayments of interest and capital, the poor countries are sending at least $20 billion a year to the industrialized world. If the falling prices of the poorer nations' agricultural commodities – coffee, sugar, cotton, etc. – are figured into the exchange, the net South-to-North flow could be as much as $60 billion a year.[3]

Throughout the 1980s, the developing countries have been spending on average between three and five per cent of their GNPs to pay their debts. More is being spent on interest payments than on either health or education. Latin America has sustained this already for eight years, most African countries for six.[4]

This transfer may be one of the most sordid international acts ever. It makes the trebling of individual giving to British famine relief charities insignificant. Band Aid's total spending on relief and development was only $140 million over five years.

Compare the sums poor countries spend on interest payments to the sums rich countries give as aid. Today, when the wealth of the United States is about 2.5 times greater than after the war, it gives as aid to the developing world 0.22 per cent of its GNP, less than any other major industrialized country. With the exception of the Scandinavians, most other Western nations have nothing to be proud of. For 20 years, aid has remained at approximately 0.33 per cent of the industrialized world's GNP, despite a widely agreed target of more than double this, 0.7 per cent of GNP.

At the same time, the quality of aid has deteriorated over the past decade, as the rich nations have used aid more and more as a way of pushing their own products. Britain gives three-fifths of its aid to other countries and two-fifths to international organizations. Between 70 and 80 per cent of the aid to other nations is tied to the purchase of British products.

THE TRAP OF NATIONAL DEBT

If he has even thought about the debt crisis, the average Northern businessman responds to the effect: "So? They borrowed the money; they mismanaged the money; they must pay back the money – just like you or me."

Not quite like you or me. National debt has different rules from personal debt, though there are similarities. The debt trap many Third World nations are now in is very similar to the debt trap many British families now find themselves in. But the Southern debt crisis is rarely explained to the average Briton, despite the fact that he or she could quickly sympathize. It is regarded as too complicated.

Advertiser David Trott makes commercials for cars, creme eggs, credit cards and, more recently, about the debt crisis. He asked the editors of such tabloid newspapers as *The Sun* and *Today* why their papers did not deal with the debt scandal. They replied that it was too complicated for their readers to understand.[5] Thus many people concerned enough to give money to save Africa's children from starving are not judged concerned enough, or smart enough, to read about the way debt is killing Third World children.

However complex, the effects of the debt crisis are too important for the subject to be left outside the understanding of ordinary people.

Eighty-five percent of the world's under-fives – nearly 500 million of them – live in the countries of the developing world. Ninety-eight per cent of all the children under five who die every year, and 99 per cent of all the mothers who die from childbirth or associated causes, die in these countries.[6]

Of course they do not all die of debt. But the debts and the economic adjustment programmes responding to that indebtedness have meant cuts in nutrition, health and education programmes. Infant mortality has already risen in parts of Latin America and sub-Saharan Africa. Already poorly nourished mothers are giving birth to more low-birthweight babies, who will in turn be more at risk during their infancy and childhood. The second half of the 1980s has seen progress in reducing this carnage slow down, stop, and in the worst cases go into reverse.

How did it happen? The higher oil prices of the 1970s were paid in dollars and other hard currencies. The oil producers stored this money in Northern banks. The banks, bursting with dollars, began a hard sell to encourage developing countries to borrow them. They did so because it seemed good business; the prices of commodities were enjoying an uncharacteristic, short-lived boom, higher than ever before in real terms. But most important was the consideration that, unlike private borrowers, nations never go bankrupt; there is simply no machinery to let this happen.

It also seemed good business to the borrowers. They were making commodity money; interest rates were low; they needed the money for development; and they would be paying back later in cheaper, inflated dollars.

Then things went wrong. In response to global slowdown in economic growth coupled with inflation, changing monetary policies in the North meant that interest rates shot up; a global recession occurred in the early 1980s; commodity prices plummeted in 1986 to their lowest levels this century; and the industries the borrowers were investing in simply were not making profits. The cheapness of the loans had encouraged borrowing nations to spend them quickly and unwisely.

Any British home-owner can understand all of this. In the early to mid-1980s, inflation was high, house prices were rising and interest rates were low. Banks and building societies were involved in a hard sell to get people to borrow; TV adverts showed paint-spattered young couples building their lives together on the basis of a real investment. Borrowing seemed good business.

Then things went wrong. Interest rates rose sharply. By the end of the decade, many people who had been encouraged by cheap loans to over-borrow found themselves – just like poor countries – heavily in debt. And the house market had slumped, making it difficult for them to get out of trouble by selling.

Most British home-owners facing increasing mortgage rates have also had some increase in income. The complaint is that every time earnings go up, the increase is swallowed by the rising mortgage rate. But Third World governments have battled against rising interest rates, declining incomes and sharply rising oil prices all at the same time. It is estimated that oil price rises alone between 1974 and 1982 account for a quarter of their debt.[7] The drop in commodity prices coincided with an increase in the prices of manufactured goods, essential imports for the developing world.

Most African debt is owed to Northern governments and international agencies such as the World Bank; most of the much larger Latin American debt is owed to private Northern banks. Governments can forgive debts, and reschedule them so that payments come due later; some have done this. But Northern governments cannot interfere too much with the debts owed to banks.

There has been a tremendous amount of tinkering by everyone involved, but the debts continue to rise. Total Third World debt stood at over $1,300 billion at the end of 1988, 90 per cent of it owed directly to institutions in industrialized nations, or indirectly through international organizations. This figure equalled half the developing nations' total GNP.[8] Nations must continue to borrow to continue to pay interest on the debts. No serious, reliable plan exists to end this downward spiral.

Latin America's debts are currently four times as large as its total annual exports. Every time the interest rate goes up by one per cent, a four per cent rise in export earnings is needed merely to keep up payments. Between the first quarter of 1988 and that of 1989, international interest rates rose by three per cent.[9]

THE EFFECTS OF "ADJUSTMENT"

When governments fall heavily into debt with little chance of repaying, the International Monetary Fund (IMF) is usually brought in to negotiate "structural adjustments" in the economy to make it more productive. A nation's ability to keep borrowing, and to get its debts rescheduled, depends on its willingness to adjust along IMF guidelines.

Adjustment programmes vary from nation to nation. But most involve devaluing over-valued local currencies, so exports are cheaper; exporting more; cutting government spending, and especially cutting the high numbers of people employed by governments; and getting rid of many of the government agencies which interfere with free-market buying and selling. Such adjustment is not evil in itself; developing nations cannot develop while following fantasy-land economic policies encouraged by the spending booms begun during the period of cheap loans.

But the adjustment programmes have affected ordinary people, especially the poorest, more than the rich (who have been part of the problem by investing so much of their money in the North rather than in their own nations). Devalued currencies have meant more expensive imports, including things like medicines and school books. Cuts in government spending and employment have generally meant rising unemployment, Free markets have generally meant rising prices.

Average incomes have fallen by 20 per cent in most of sub-Saharan Africa, and by 10 per cent in Latin America. Even within these averages the extent of some people's suffering is hidden: it is thought that in some cities family incomes have been halved.[10]

The need to increase exports also tempts governments to over-cultivate fields and over-use water to produce commodities.

Indebted nations with forests have cut into them heavily to earn hard currency. Though it is hard to prove, there are widespread fears that rising unemployment is forcing more people on to the land – often poor and marginal land – and thus accelerating environmental degradation. All of this robs future generations of resources for development.

Less government spending has meant drastic cuts in public services, and since the cuts prescribed are large and the budgets already small, even the most farsighted governments must cut services essential to children's welfare. But the question of where to cut is left largely to governments, and many have made the wrong choices, from the point of view of children, the environment and the future.

Military budgets have suffered little. In 1990, half of all Third World government spending was either on debts or on the military, these governments as a whole spending about $125 billion a year on the military. UNICEF reckons that about a week of this spending, invested each year in buying and using simple medicines and vaccines and anti-diarrhoeal solutions, on top of what is already being invested in this manner, could eliminate most of the deaths of children under five in the developing world each year.[11] Those children die through choices made by their own and Northern governments, not by necessity.

Neither has the little money remaining from debt servicing and military spending been spent well. Money has consistently gone to prestige projects like big city hospitals and universities, rather than to rural health centres and primary schools, both of which would be much more cost-effective in terms of providing health and education. Despite the recognition that investing in women's health and education is the best way of improving their children's health and education, women and girls continue to suffer disproportionately.

The many principles embodied in the 1989 UN Convention on the Rights of the Child add up to the single principle that providing for the basic needs of children, psychological and social as well as physical, should be the first priority, yet cuts in spending on children's nutritional needs, health and education show that children are among the lowest governmental priorities.

Statistics showing the full extent of the suffering and its longer term effects are only starting to appear, but already it is clear that a large part of the developing world is moving backwards, and that children are bearing the brunt of the reversal.[12]

Spending on managing environmental resources such as topsoil, water, forests and clean air, which will have a profound effect on these children's futures, has also been cut savagely. Both Africa and Latin America lack not only the cash but the experts even to monitor what is happening to their environments, much less to do anything about the damage.

"Crisis" does not mean disaster; it means turning-point, and comes from the Greek word for "decision". But there has been no turning point in the debt crisis, no thoughtful decisions. There are no obvious solutions. Many Third World environmentalists actually oppose writing off the debts, arguing that it would only encourage their governments to accelerate the plunder of the environment. But for the sake of children, of the environment, and of the future, solutions must be found.

As the World Commission on Environment and Development wrote: "a necessary sense of urgency is lacking. [Solutions] must incorporate the legitimate interests of creditors and debtors and represent a fairer sharing of the burden of resolving the debt crisis".[13]

CHILDREN AND FOOD

The poorer a family, the more of its income goes on food and other basic necessities. The poorest families spend up to three-quarters of their total budget on food. With substantial cuts in income and rises in basic food prices, they have no choice but to feed themselves and their children inadequately.

More than 150 million children under five in the developing world today suffer from malnutrition; undoubtedly their numbers have been increased by adjustment programmes and cuts in government food subsidies.[14] In Zambia, the incidence of malnutrition as a factor contributing to the deaths of children under 14 rose from 27 per cent in 1978 to 43 per cent in 1982.[15] Today, over 40 per cent of all Zambian children between the ages of two and five suffer stunted growth, a condition associated with long-term malnutrition.[16]

Malnutrition has surfaced rapidly under adjustment programmes. In Chile in 1983, in the midst of a worsening economic crisis, food supplement programmes were cut. The amount of free milk given out was reduced by 31 per cent between 1982 and 1983, and additional food distribution to malnourished children by ten per cent.

The effects were seen very quickly. Malnutrition figures for the months of February, March and April of 1983 showed a stagnation in the nutritional status of children younger than five months old, and a deterioration in that of children 6–23 months old. The yearly figures confirmed this downward trend.[17]

CHILDREN AND HEALTH

During the 1980s, health spending per head dropped by over 50 per cent in the poorest 37 countries of Africa and Latin America, and infant mortality has already risen in some of them.[18]

Three-quarters of the money governments spend on health services in the developing world is still spent on hospitals and on expensive curative medicine for a tiny proportion of the population. The poor – the majority in most developing nations – cannot afford to see the insides of these hospitals.

The debt crisis and adjustment programmes have further tilted the balance away from the needs of the neediest. Large city hospitals have generally not had their budgets reduced, but hundreds of rural health clinics have been closed down or suffered cuts in staff and supplies. In countries like Botswana and Jamaica, training of community health workers has had to be suspended.

The value of primary health care, based on health workers less well trained than doctors working in the countryside and slums, has been proved in those countries where it has been given an increased share of the national health budget.

Pakistan reduced its hospital budget to 41 per cent of its overall health budget, with dramatic results. Between 1984 and 1988, Pakistan increased its immunization coverage from five to seventy-five percent of all children. According to Pakistan's former minister of finance and development planning, the entire cost of the immunization campaign was paid by postponing for five years the decision to build one expensive urban hospital.[19]

CHILDREN AND EDUCATION

In the 37 poorest countries of the world, spending per head on education has been cut by a quarter during the 1980s, and primary schools have suffered most.

As Frederico Mayer, the director general of UNESCO put it, "The greatest damage seems to have been done at the very foundation of the educational pyramid, that is in primary education, and in basic literacy for adults and out-of-school youth." More than half of government spending on education in the Third World goes on secondary and higher education, which are only enjoyed by 30 per cent of the population.[20]

Foreign aid reinforces this, with only one per cent of all aid going to primary schools. One hundred children can graduate from primary school for the cost of one graduate from a university. And those 100 will be literate, numerate citizens, while the university graduate may have to go abroad to find a job.

The importance of primary school is obvious: if a child cannot read, write or deal with basic figures, he or she can hardly be educated further. In many countries, including Bangladesh, Guyana, Madagascar and Mexico, enrolment rates at primary schools have declined over the past few years, and dropout rates are increasing, particularly for girls. For most of sub-Saharan Africa, figures are only available up until the end of 1986, but the proportion of all 6–11 year olds enrolled at primary school is already falling.[21]

Sending children to school has to be weighed against their possible contribution to the family income by working – the tighter the budget, the more difficult the decision, even when education is free. Among the poorest, other school expenses, like uniforms and fares, can be enough to keep a child at home; hunger and malnutrition-related diseases can be enough to prevent children either from staying at school or from fulfilling their potential. As cuts in expenditure make teachers and materials scarce, and the quality of education can be seen to deteriorate, already hard-pressed parents will be even less likely to make the sacrifice.

Many World Bank studies have confirmed that investment in education is money well spent, particularly in the poorest countries. In Africa for instance, rates of return for investment in education can be very high, between 10 and 20 per cent, compared to an average return on investment in Africa of seven per cent.[22]

Educating girls may be an even sounder investment over the long term, yet cuts in family income and government services are more likely to mean that girls remain uneducated. With no state child care, it is the older sisters who are kept at home to look after younger brothers and sisters when both parents need to go out to work to keep the family alive. Studies have shown that maternal education is the biggest single influence on the fall in infant mortality rates. Women who are literate are more likely to have fewer and better spaced children, each of whom will have a better chance of surviving to adulthood.

More than two-thirds of those children who never go to school or who drop out before completing primary school are girls. At the beginning of the 1990s, a girl born in South Asia or the Middle East had less than a one in three chance of finishing primary school.[23]

HUNGRY, SICK AND UNEDUCATED CHILDREN

In the real world, the effects of malnutrition, ill health, bad housing and lack of education cannot be separated one from another. Attempts to improve the lot of children by spending in one area are likely to be thwarted by inadequacies elsewhere.

Despite relative and absolute increases in the budget for primary school education in Chile between 1974 and 1983, the number of children finishing primary school failed to rise. In 1984, at least 12 per cent of poor households in parts of Santiago had primary school aged children who were not going to school. It is hard to tell precisely why children drop out of school or fail to attend regularly, but it was assumed that a drop in family income and the poor nutrition of young children were both significant.[24]

Schooling and health care both act as multipliers on other forms of investment. Well-fed children learn better and grow into healthy, educated adult earners. Improvements in health and education are a cause as well as a consequence of overall development in any nation.

But where primary education and health care suffer first and most, this cycle is viciously reversed. The effects are felt immediately by children, and the long-term effects may be even more devastating.

Other effects of the debt crisis are harder to establish statistically but are beginning to make themselves felt. More desperate families mean more abandoned children living on the streets, more children at work and suffering from accidents there, more juvenile delinquency and more children involved in drug abuse.

The 1989 UN Convention on the Rights of the Child has specific provisions protecting children from each of these threats. In addition to minimum standards of protection for children's survival and welfare, the Convention calls on governments to protect their children against physical or sexual abuse, and exploitation at work or at war.

CHILDREN AT WORK

In the poorest communities of developing countries, child labour is common both in cities and in the countryside, in the industrial and service sectors, on plantations and in sweatshops.

In 1982, it was estimated that there were at least 145 million child workers worldwide between the ages of 10 and 14, although some estimates offer a much higher figure. Official figures do not take account of the large numbers of children employed as casual workers. In Brazil, nearly a third of all domestic workers are children, some of whom are taken on at three or four years old.[25] Even in Italy, an estimated 1.5 million children work illegally in small factories or as outworkers doing piecework from their own homes.[26]

In most societies, all children work at some time or other, even if it is only during weekends or school holidays. Work itself is not necessarily the problem. For the Anti-Slavery Society, child work becomes child labour when children are made to work in conditions which impede their normal physical, social or psychological development.

The exploitation suffered by children in the workplace is a more acute version of the inequity endured by many adult workers, but because of their immaturity, children are less likely than adults to assert themselves; too young to join unions, they have no collective organization, and if they do have any rights, they are unaware of them. Children are often made to work long hours in dangerous conditions for less pay than adults. Cheap child labour is usually good business, particularly for small enterprises desperate to keep their costs down to remain competitive. Inadequate safety measures for child workers cause a great deal of death and injury which is never officially recorded. As well as physical suffering, children suffer emotionally and intellectually when deprived of the stimuli of play and school.[27]

In poor communities, child labour is both a cause and an effect of poverty. However hard the work, families feel fortunate when a child can earn the difference between enough to eat and hunger. But in the long term, child labour is part of the problem. For every child worker, there is an unemployed adult, and for every unemployed adult there is another hungry family. As Alan Whittaker of the Anti-Slavery Society puts it, "The sad paradox at the heart of child labour is that it perpetuates poverty because it is a cheaper alternative to already cheap adult labour . . . Child labour prevents the growth of organised trade unionism and maintains a Victorian-type boss's capitalism. It stunts dignity and perpetuates powerlessness."[28]

The conventional school system is not appropriate for all working children, including runaway and abandoned street children. A highly academic curriculum, inflexible hours and

Overleaf Poor children make toy cars for rich children in a sweatshop in Dacca, Bangladesh. *Mark Edwards/Still Pictures*

expensive textbooks and materials all keep poor children away. Many experts argue that education policies must be more flexible to allow children to be educated at the same time as they contribute towards the survival of the family. This is hardly a new idea. Until the second quarter of this century, holiday periods in schools in southern parts of the United States were built around the cotton harvesting season.

In some countries, attendance at school is combined with seasonal or part-time work, and some voluntary organizations have experimented with setting up pavement schools for children working on the streets. The Underprivileged Children's Education Programme works with 16,000 street children in Bangladesh and Nepal. Every day, the children attend one of four two- to three-hour school shifts, according to their work routine.[29]

Child economic exploitation reaches its lowest form in child prostitution, child pornography, sex tourism and the sale of children for sexual purposes. Contrary to popular prejudice, commercial sexual exploitation is not restricted to South-east Asia and other parts of the developing world. An estimated 5,000 boys and 3,000 girls work as prostitutes in Paris, and it is thought there may be as many as 300,000 boy prostitutes in the United States.[30]

The prevalence of prostitution in South-east Asia is linked to the US military presence in the area during the Korean and Vietnamese wars. Providing sexual services was seen as an essential part of looking after the recreational needs of military personnel and designated areas where prostitutes, including children, were available were built in Vietnam, Thailand, the Phillippines, Korea and Taiwan. When the war ended in Vietnam so did most prostitution, but it remains widespread in other parts of South-east Asia.

More recently, the tourist industry has encouraged commercial sexual exploitation in South-east Asia. Some European, Japanese and US travel agents offer package holidays which cater specifically for sexual gratification. Many prostitutes servicing tourists are under the age of 18, and many may be under 16. Publications have started to provide information on how to find children for sex.

There are thought to be 30,000 prostitutes in Bangkok under the age of 16 and 2,000 male prostitutes between the ages of 7 and 17 in Sri Lanka. Many of the Thai children are recruited for a fee from parents living in poor rural areas on the understanding that they will be given good jobs in the city. Child prostitutes in Southern Thailand work 12-hour shifts during which they have sex with between five and ten clients. They are paid 25–30 cents per shift.[31]

STREET CHILDREN

Street children are a common sight in most of the towns and cities of the developing world, particularly in South America and Asia. No one really knows how many of them there are; estimates vary between 100 million and 200 million.

Commentators have identified three distinct groups of street children. The largest, accounting for perhaps 60 per cent of all street children, are "children on the street" or children who work on the streets and largely survive there, but who still live with their families. The second group, about 33 per cent, are "children of the street", who view the street as the only home they have and their only source of shelter, food and companions. They have left their families to live independently, although a lot may have left home unwillingly and wish to go back. The remaining seven per cent of street children are abandoned children who live like the others, but with all family ties severed.[32]

The fate of all of these children is intimately bound up with the economic plights of their families and their nations. Putting a child out to work on the streets may be the only option left to a family with children it is unable to feed. During the 1980s, millions of the poorest families have been pushed towards this option.

Children living on the streets automatically commit offences since their life-style is itself illegal. Many children end up in inappropriate institutions or adult prisons. Where resources are lacking, children who are held in care for many different reasons – criminal acts, absence of identity papers, vagrancy, parental abandonment or abuse – may be held together with adult criminals in institutions where punishment or correction, rather than child welfare, is the primary aim.[33]

CHILDREN AT WAR

In the First World War, civilians accounted for just five per cent of the total number of victims; in the Second World War, 50 per cent; and in the Vietnam war, between 80 and 90 per cent. Due to changing weapons and tactics, the civilian death toll has risen throughout the twentieth century to about three-quarters of all deaths. Despite a 1974 UN declaration which condemns attacks on civilian lives, civilians rather than soldiers have been primary targets as well as accidental victims of war in recent years.[34]

Deaths in wars represent only a fraction of overall losses. Huge numbers of people are injured, and material losses in income, trade and bombed homes and schools depress living standards for decades afterwards.

Disruptions to welfare services during and after times of war hit the neediest children hardest. Both governments and resistance fighters have always found, and continue to find, uses for children during war. Their agility and loyalty suit them for work as intelligence gatherers and messengers. Children are enlisted as fighters, arrested, and executed or tortured to death throughout the world.

Numerous Amnesty International reports have revealed that children's rights are being violated for political purposes in nations where governments wage undeclared war against large numbers of their citizens. Babies and children in Iraq have been kept in detention with their parents and used as tools to force their parents to confess to alleged political offences. Children as young as six have been arrested and detained without charge or trial in lieu of relatives being sought by the authorities.

Numerous children and young people are among those that have "disappeared" in Iraq. The term "disappearance" is used by Amnesty International whenever there are reasonable grounds to believe that a person has been taken into custody by the authorities, or with their connivance, and the authorities deny that the victim is in their custody. The "disappeared" are often victims of violent arrests or torture; many die in custody.[35]

International law makes it illegal for children under 15 to fight in wars, but children younger than this are fighting in resistance groups and civil wars all over the world, including Northern Ireland and Beirut. Many fought for Iran in its war against Iraq.

During negotiations for the 1989 UN Convention on the Rights of the Child, many delegates argued to raise the age at which children can fight to over 15, but this was resisted. The Convention says that states which have agreed to it "shall take all feasible measures to ensure that persons who have not attained the age of 15 years do not take a direct part in hostilities". Of the 127 wars this century, the vast majority have been in the developing world.[36]

POVERTY AMID WEALTH

Children are much better off in the wealthy Northern nations. Their parents not only have more political and economic power, but they balance the precarious act of raising a family over a safety net constructed of sick pay, unemployment benefits, child benefits, free education and health care.

But there is alarming evidence that most European and North American governments find it hard to fit the needs of children, the basis of future national prosperity, into their present policies focused on wealth production. Governments which lack the excuse of poverty but still cannot provide safety nets for the poorest children, often large numbers of children, have no hope of achieving sustainable development.

"Poverty is a relative as well as an absolute concept. It exists, even in a relatively rich Western society, if people are denied access to what is generally regarded as a reasonable standard and quality of life in that society", the Church of England wrote in its report on inner cities in 1985.[37]

In Britain and the United States the number of homeless families has doubled during the last decade, despite ten years of steady economic growth. In 1990, the US government compared the status of US children against those in 11 other industrialized nations. The only nation having a higher infant death rate than the US was the USSR – 10 deaths before age one for every 1,000 births in the US, versus 25 per 1,000 in the USSR. Jointly the United States and Australia topped the tables with 17 per cent of their children living in poverty. "The dead babies, the murders, the child poverty are a haemorrhage on human resources and the American spirit", said Congressman George Miller, whose committee issued the study.[38]

Most welfare problems – bad housing, poor diet, ill-health and illiteracy – spring from poverty. In Britain, at the end of the 1980s the "average" person owned slightly more of the national wealth than he or she did 20 years before, largely because of the rise in home-ownership.[39] But although most people's standard of living has risen in the last ten years, the number of poor and low-paid has also risen substantially, and the gap between the wealth and income of the richest and poorest has widened. The top ten per cent of households had an average rise in real income of 18 per cent between 1979 and 1985, compared with only six per cent for the bottom ten per cent.[40]

"Poverty" is an abstract word, but the effects are specific and regular. Britain has no official poverty line, but it is generally taken to be the level of Income Support (formerly Supplementary Benefit) paid by the government to the unemployed. In 1985 there were 15.4 million people living either in poverty or on the margins.[41]

According to one study, in four out of ten poor families, children have to go without meals, are ill-protected against the cold and normally wear second-hand clothes, for lack of money.[42] Many poor and low-income families find it difficult to meet family food budgets, particularly fresh fruit and vegetables all year round.[43]

Unemployment and increases in divorce and separation resulting in more one-parent families are the two biggest factors in the rise in poverty during the 1980s. In 1985, nearly two-thirds of the children in single-parent families were living on or below the Income Support level, compared to 12 per cent of children in two-parent families.

One major problem facing single parents is finding work that fits in with childcare, especially as few local authorities provide such help. Many single mothers depend on Income Support. Single parents used to be able to deduct childcare costs from earnings when claiming benefits, but under current rules they can no longer do this. So many single parents, particularly those on low pay, may be better off financially by not having a job.

As with poverty, there is no official definition of "low-paid". The Low Pay Unit, an independent pressure group, defines it as earnings below two-thirds of the average earnings for adult men. In April 1988 this low-pay benchmark was £143.67 per week, or £3.80 per hour for part-time workers. In cash terms this is slightly less than the Council of Europe's "decency threshold" figure.

In April 1988 more than nine million people in Britain – about 45 per cent of the workforce, compared to 36 per cent in 1979 – were earning less than the Low Pay Unit's threshold rate. More than five million of these were working full-time and four million part-time. The majority were women, more than six million in all, working full- or part-time, many with young children.[44]

IN SICKNESS AND IN WEALTH

In Britain, in 1985 the child of an unskilled worker was more than twice as likely to die before its first birthday as a baby born to a professional worker.[45] In America, a newborn black baby bears the same double risk compared to a white baby.[46]

Being poor makes you sick; this was the basic message of the Black Report, produced by a working group set up by the government in 1977 and published in 1980.[47] "Material deprivation" was found to be the major factor in the inferior health records of the poor, particularly partly skilled and unskilled manual workers who made up more than a quarter of the population. Biological, cultural and life-style factors were also important, but less so than material deprivation.

The rich–poor health gap has not been narrowing or even staying the same; it has been widening, according to a 1987 report by the Health Education Council. Again, although genes and life-style were shown to play a part, the key cause of inequalities in health and early death was found to be "material deprivation": bad housing, bad environmental conditions and facilities, bad working conditions, unemployment, and above all, inadequate income.[48]

Much of the pattern of inequality is established in childhood, infancy or even before. Low-birthweight babies are more likely to be born to the partners of unskilled and semi-skilled workers, hence the increased rates of child illness and death in these groups. Later on, particularly during the first five years of life, these children are more likely to die from the two major causes of all child deaths: accidents and respiratory disease. Poor parents "simply lack the means to provide their children with as high a level of protection as that which is found in the average middle-class home. This can mean both material and non-material resources".[49]

The Health Education Council's report concludes that "there is considerable evidence that material deprivation affects physical development in young children and that ill health contracted in childhood can dog an individual for life". Poor children are more likely to suffer from recurring bouts of illness and less likely to be adequately treated for them.

HOUSING

From 1961 to 1988, home ownership in Britain doubled. But during the 1980s, homelessness in Britain doubled. Homelessness has been growing in Britain since the beginning of the 1970s, swollen by the rise in unemployment, compounded in the 1980s by a decreasing stock of public housing. Of all areas of state spending, housing has been hardest hit, and the situation is now acute. In 1988, the Audit Commission warned that inner city councils have become so overstretched by demands from the homeless that they will soon be unable to cope. Families with young children and young single people are suffering particularly.

In 1989, more than 150,000 young people were likely to be homeless at any one time in Britain, according to the housing charity Shelter. There are no figures available on exactly how many of the young homeless are under 18, but it is widely recognized that increasing numbers of 16 and 17 year olds are facing the terrifying insecurity of homelessness and temporary housing.

Since changes in benefit regulations in 1988, 16 and 17 year olds can no longer claim benefits except in very special cases. Without a job or a place on an official youth training scheme, and unable to live with their parents for whatever reasons, some teenagers find themselves without any source of income. Although local authorities in Britain have a statutory duty to provide accommodation for all those in priority need in their area, in reality this does not usually include young, single people, who are given lower priority than families.

Once homeless, teenagers find it more and more difficult to get or hold down a job or training scheme place. While home owning has become more of a possibility for the majority in the 1980s, chances of finding any kind of accommodation, even temporary, have become slimmer for the minority. At the end of the decade, it was more difficult for single young people to find somewhere decent and affordable to live than at any time since the 1950s, according to Shelter.[50]

The number of homeless households accepted by local authorities in England and Wales in 1988 was 135,000, a four per cent rise on 1987, and more than double the 1979 total. Nearly all of these households were classified as "priority need", and two-thirds of these were families with children.[51] Once accepted by the council, families often end up in temporary accommodation, bed and breakfast hotels, hostels (including women's refuges) and short-life tenancies.

Living without a home, albeit beneath a roof, in substandard temporary accommodation puts enormous physical and psychological pressures on families and children. Surveys of people living in "homeless accommodation" have found that many families cannot afford to eat properly. Nearly a third of women interviewed said they went without food, and one in ten mothers said they could not afford enough food to feed their children.[52]

Many families must be housed outside their own boroughs, which makes it more difficult for them to use medical, social and educational services. In London in 1988, there were about 12,000 children in bed-and-breakfast accommodation who found it hard to enrol at, or succeed in, local schools. The Audit Commission has calculated that the cost of building a council home is only £7,400 a year, while that of keeping a family in bed and breakfast is £11,315 a year. In 1988, London's local authorities spent £100 million on bed and breakfast, compared with £12 million in 1985.[53]

EDUCATION: "THE POOR AND THE SHODDY"

More than half of Britain's teenagers leave full-time education at the age of 16. Full-time education and training combined still only account for 50 per cent of British 16 year olds compared to 69 per cent in West Germany, 92 per cent in Japan, and 94 per cent in the United States. Only 20 per cent of all Britain's teenagers leave school with even one "A" level.[54]

Over the past ten years, there has been very little overall improvement in standards of achievement in most areas of education. A third of British school leavers have no useful qualification to show for at least 11 years' full-time education. On average, British children are at least two years behind the Japanese in maths, less likely to learn a foreign language than the French, Germans or Scandinavians, and generally have fewer and lower educational qualifications than the young people of most other European Community countries.[55]

In a 1990 report on standards in education by Her Majesty's Senior Chief Inspector of Schools, it was pointed out that some 30 per cent of teaching in schools and 20 per cent of further-education teaching was judged to be poor or very poor. On top of this, less able pupils are more likely to "experience the poor and the shoddy" than the average pupil. This further disadvantaging of the already disadvantaged is described as "a worryingly persistent feature of English education at all levels".[56]

Poor education is showing up in job skill shortages. "The UK is seemingly stuck in a vicious circle", Lloyds Bank warned in 1990. "Low average skill levels in the workforce encourage many sectors of industry to adopt low-technology production processes, which in turn require low levels of training, exacerbating the relative skill shortage".[57]

According to the Inspector's report, differences between learning styles and types of assessment for children studying between the ages of 16 and 18 make those years "a jungle in which talent and ability are lost". Without coherent national guidance, the development of post-16 education and training will continue to be "slow, hesitant, contradictory and counter-productive".[58]

Britain has a long history of selling education short. A Royal Commission in 1868 warned that "our industrial classes have not even that basis of sound general education on which alone technical education can rest". In 1956, a White Paper on Technical Education warned that Britain was in danger of being left behind compared to the United States and the rest of Western Europe in terms of technical and scientific manpower. The Plowden Report in 1967 pointed out Britain's failure to "provide the educational background necessary to support an economy which needs fewer and fewer unskilled workers and increasing numbers of skilled and adaptable people".[59]

Educational achievement in Britain is still closely related to class, wealth and where one lives. Twenty-five per cent of the children of men in professional occupations go to university, compared to 12 per cent of the children of non-manual groups and two per cent of manual workers.[60] In the year 1985–86, 51 per cent of 16 year olds stayed on in full-time education in the South-east, compared to only 39.8 per cent in the poorer North.[61]

CHILDCARE

Between 1971 and 1988, the number of women working in Britain rose by 27 per cent to almost 12 million. Women now account for 43 per cent of the workforce, compared to 37 per cent in 1971, and their numbers are likely to rise.[62]

With a growing demand for women to replenish the workforce and more and more women choosing to return to work before and during their children's school years, new arrangements must be found for minding their children. Britain, along with Ireland, The Netherlands and Luxembourg, has the worst record on childcare of all European countries, providing publicly-subsidized care for only one per cent of under-threes and the equivalent of full-time care for only about 25 per cent of children between three and school age. In contrast, 85 per cent of pre-primary school children in Italy, France and Belgium are provided with a minimum of six hours' care per day.[63]

The 1989 UN Convention on the Rights of the Child gives the children of working parents the right "to benefit from child care services and facilities". Guaranteeing this right requires not just more facilities, but methods to give poorer parents access to those facilities.

But these issues reach beyond simple labour force concerns into themes equally crucial to society: equality, child welfare and family life. All the evidence shows that women working outside the home, full-time or part-time, continue to do most of the work they have always done inside the home, including childcare. Thus there is a danger that more "flexibility" in women's careers may mean nothing more than the flexibility to do two jobs.

It is men's lives which are inflexible, to the detriment of family life and children's well-being. Only when men are freed from traditional full-time work patterns, only when they take on more responsibility for childcare and the home, can women make any significant progress in employment and careers.

The UN Convention on the Rights of the Child recognizes the principle that both parents have joint primary responsibility for bringing up their children and that the state should support them in this task. There is little evidence to date to show this is happening in Britain.

CHILD ABUSE

The whole concept of child abuse is relatively modern. Dr Henry Kempe, a paediatrician, first wrote of what he called "the battered child syndrome" in 1962.[64] Since then research has widened the definition of child abuse to include emotional, sexual and psychological as well as physical abuse and neglect.

Those who work with abused children acknowledge that actual abuse far exceeds the number of cases recorded officially. In England and Wales in 1987, the National Society for the Prevention of Cruelty to Children (NSPCC) recorded over 7,000 cases of sexual abuse and over 8,000 cases of physical abuse. These figures do not include other recognized categories of abuse such as physical neglect (often starvation), failure to thrive and emotional abuse such as rejecting or terrorizing a child.[65]

In 1988 more than 40,000 children were on Child Protection registers in England and Wales, a rate of about 3.5 children per 1000 in the population under 18 at risk from some sort of abuse from their parents or guardians.[66] Child abuse of all forms appears to be increasing, but it is difficult to tell how much of this is due to an increase in professional awareness of the dangers children face. In 1987, the "Cleveland affair", in which a local authority separated scores of parents and children suspected of being sexually abused, brought sexual abuse at least partially out of the closet.

Some researchers have argued that the stresses of urban industrial society, the demise of the extended family, family isolation and the lack of wider family support contribute to the risk of child abuse. This seems to be true in the Third World. In sub-Saharan Africa, abuse and neglect are still very rare, but the incidence appears to be rising, particularly where the emotional and physical stress on parents is increasing with the collapse of traditional social patterns.[67]

The reasons why children in general and some children in particular are abused by adults are poorly understood. Explanations are incomplete and often contradictory. But cultural values seem to be important. Studies in the Third World show that where children are valued for a specific reason, as income generators or the perpetuators of the family, they are unlikely to be abused. But in cultures which attach less value to girls than to boys, girls are more likely to be abused.

Thus sexual abuse of children in Northern countries may have roots in Northern cultures which exploit the sexuality of the less powerful – women and children – in everything from hard-core pornography to page three pin-ups and titillating advertisements. In the 1990s, children are no longer sent up chimneys, but they do feature in child pornography in many countries, such as The Netherlands which has no law against pornography. Many people who express genuine horror at individual cases of child abuse can live with the more pervasive manifestations of the same root abuse of power.

Third World studies also show that child abuse and neglect is minimized in many cultures because children are not judged to be responsible for their actions until the age of seven or eight. Dr Jill Korbin points out that this is "in direct opposition to the age-inappropriate expectations of very small children and infants that are so often implicated in child abuse and neglect in Western nations".[68]

THE CHILD CONSUMER: CATCHING THEM YOUNG

Children care deeply about the environment. British company directors surveyed in 1989 said the main reason for their own growing concern for the environment was the regular haranguing they were getting from their children on the state of the world. Opinion polls, the stacks of new books telling children how to be green and the sacks of mail from schoolchildren to environmental groups all reflect this youthful concern for the state of the world.

But there is a stronger tide in the North running against this concern. Advertisers are teaching children to want, to want more and more, and to define their own well-being by the quantity and the brand names of their possessions. Some of this advertising is the most cynical commercial exploitation of the innocent; some of it is for products which kill.

Worldwide, 80 per cent of advertisements aimed at children are selling toys, cereals (mainly sugar-coated), sweets and fast-food restaurants. A monitoring of British late afternoon and Saturday morning children's television found that more than half of all the advertisements were for food, and four-fifths of these were for junk foods high in sugars and fats: sugar-coated cereals, confectionery, soft drinks, snacks and fast food.[69]

The advertising of cigarettes – the only legally sold product which, used as directed, can be fatal – is restricted in the North. So the tobacco industry invests heavily in the sponsorship of sports and arts events, much of which materializes as television advertising. The sports connection is especially compelling for teenagers.

Most smokers take up smoking before the age of 18. "The matter is largely settled by the age of 20: if a person is still a non-smoker by this age he is unlikely to take it up", according to the British Royal College of Physicians. Even where cigarette advertising is severely restricted, children are still accessible victims of advertising. Children and young people tend to smoke the brands that are promoted most heavily, according to the World Health Organization (WHO).[70]

In 1989, US president Bush asked for $7.9 billion for a war against drugs. In 1988, US hospitals reported 3,308 deaths attributed to cocaine. They also reported 390,000 deaths in some way attributable to the use of tobacco, the growing of which receives government subsidies.[71]

As their customers die off, cigarette companies become ever more ingenious in attracting new ones. R.J. Reynolds, one of the world's biggest tobacco companies, has been accused of inducing children to smoke by using cartoons of a high-living, street-wise camel as the centrepiece for their advertisements for Camel cigarettes in the United States.[72]

R.J. Reynolds has also been developing a marketing campaign for a new cigarette brand directed at poorly educated white women aged between 18 and 24. "Project Virile Female" aims to sell "Dakota" cigarettes to women whose main aspiration is "to have an ongoing relationship with a man". The company planned to sponsor male strip shows and produce "hunk" calendars. Any product aimed at 18-year-old women is almost by definition aimed at girls younger than 18 who want to appear sophisticated.[73]

Many young women have babies, and the uneducated are more likely to smoke during pregnancy. Babies born to women who smoke tend to be about 200 grams lighter than those born to comparable non-smokers. Smoking women inhale various poisons, including nicotine and carbon monoxide, which reduce the amount of oxygen reaching the foetus and restrict its growth. The chances of still-birth or death of an infant within the first week of life increase directly with the number of cigarettes smoked during pregnancy.[74] Babies and young children are also particularly at risk from respiratory or ear infections through "passive smoking", which is thought to cause up to 5,000 deaths in the United States and 1,000 deaths in Britain every year.[75]

Although smoking is declining in Europe and is stagnant in other developed countries, it is increasing in all regions of the developing world, particularly in Africa.[76] The Indian tobacco company Golden Tobacco recently launched a brand of cigarettes, called "Ms", aimed at young women. In many developing countries cigarette packets have no health warnings; tar contents in some brands are much higher than in identical brands sold in the North, and much cigarette advertising sells smoking on the grounds of status and success, an approach long outlawed in the North. Tobacco companies see tremendous room for expansion in the South, as the average Third World smoker consumes about 300 cigarettes a year, compared to the 2,500 per year of the northern smoker.[77]

At the moment far fewer young women smoke in Africa and Asia than young men. This was the case in much of the developed world up to the Second World War, when smoking began to increase rapidly among young women. In Britain today, 15 per cent of 15-year-old boys smoke compared to 22 per cent of girls of the same age.[78]

As Margaret Thatcher said: "We need to emphasize particularly to young children the message that smoking kills . . . Among an average of one thousand young adults who smoke cigarettes regularly about six will be killed on the road – that it bad enough – but about 250 will be killed before their time by tobacco".[79]

But teaching Northern children to *consume* is as dangerous for the planet as teaching one child to consume cigarettes is for that child. The quarter of the planet's population which lives in the North already consumes three-quarters of the planet's resources. This rate of consumption, the debt crisis and low commodity prices encourage the South to export its natural resources at bargain basement rates.

Since the 1960s, advertisers have increasingly sought to exploit the child and youth markets. They want to hook them on brand names at an impressionable age. As the US girls' magazine

Seventeen put it to its advertisers, "If you miss her then, you miss her for ever. She's at that receptive age when her looks, her tastes and brand loyalties are being established . . . Reach a girl in her *Seventeen* years and she may be yours for life . . .".[80]

In 1985, Britain's 7 to 17 year olds spent £320 million on clothes, £198 million on confectionery, £179 million on records and tapes, and £110 million each on crisps and soft drinks. Market researchers confirm that the young consumers of the 1980s are "aware of brands and status items before they can read".[81] Unsustainable and inequitable patterns of consumption are likely to be maintained and increased by today's unprecedented advertising and promotion to children. Advertisers sell dreams, and children's dreams are easily preyed upon. They are more deserving of society's protection.

The forces at work today producing a new generation of wanters in the rich nations may be the most unsustainable of all the unsustainable development patterns at work on the face of the planet.

3
POLLUTING POSTERITY

> Too many political leaders give more attention to the obvious costs of action than to the concealed costs of inaction.
>
> *Prince Charles, HRH The Prince of Wales*
>
> Most of the costs of dumping poisons into air, land and water will be paid by future generations: first, because the Earth's recycling system moves slowly; second, because the future generation now present in the sperm and ovum of living people is being damaged as the interior of the human body gradually assumes the polluted state of the external environment.
>
> *Rosalie Bertell, health expert*

The scene is a classroom in the heavily industrialized, heavily polluted Silesia region of southern Poland. The teacher is conducting a lesson for her 11- to 13-year-old students on the environment. In the back of the room, a sociologist who specializes in children's reactions to their social and physical environment is quietly taking notes. The lesson has hardly begun before the students take over:

Teacher: "Our lesson is about life in Silesia. We are going to paint a picture of our environment, our surroundings. What should we put in the picture?"
Students (all at once): "Coal mines! Factories! Steel works! Houses! Slag heaps! Streets!"
Teacher: "Hmm. Lots of streets – because all the towns are right next to each other. What else should be in the picture?"
Girl: "Lots of chimneys, fumes, and everything should be in a grey colour."

Teacher: "Yes, grey. Wha . . . ?"
Boy: "There is a lot of dust in the air and people can get collier's lung disease. That's when the dust gets in the lungs and the lungs produce a second tissue over affected places."
Teacher: "Well said. But how do you know so much about it?"
Boy: "I read a lot."
Teacher: "Do you know any children or grown-ups who suffer from collier's lung disease or bronchitis?"
Another boy: "My parents have got pneumonia and bronchitis."
Previous boy: "My father's got collier's lung disease!"
Teacher: "Your father? So that's why you know so much about it. But the environment also affects babies, even before they are born. I was reading a book which says that the death rate for newborn babies is much higher in the Katowice area [the industrial heart of Silesia] than in any other region of Poland. How does that happen?"
Girl: "Mothers breathe the bad air and it gets into their bodies and surrounds the baby."
Teacher: "Yes, and I read that the highest death rate is in Zabrze, followed by Ruda Slaska. So Ruda Slaska, where we live, has the second worst death rate for newborn babies in the country. What do you think of that?"
Boy: "That is horrible!"
Teacher: "Yes, one word 'horrible'. I would like . . . Iwonka, you wanted to say something?"
Iwonka: "My mum works at the steel works at Pokoj. She was expecting a baby, but it was still-born."
Teacher: "Is your mum still working there?"
Iwonka: "Yes."
Teacher: "So she lost her baby but still works in the plant. Were you sad when you learned about it? What did you think the reason was?"
Iwonka: "There are gases and fumes and my mum was inhaling them and the baby was already poisoned in mum's tummy."
Teacher: "Did you think like that then or did your mum tell you?"
Iwonka: "My mum told me."

Iwonka has begun to cry steadily and the teacher goes to hug her. The lesson slowly resumes. The students talk about how their mothers must wash the windows every day so the family can see out. They discuss the clean air in the mountains, where some of them go on holiday for a week once every few years. Eventually the lesson ends and students file out. The sociologist, Dr Krzysztof Stadler, talks to the teacher about what they have both heard:

> These children have a high level of knowledge about the dangers in their environment. But I have done research among children aged 14 and 15 in this region, and have found that this level of concern unfortunately disappears or becomes weaker as children grow older. Older children develop a dulled sense of menace. They become used to the idea of living in a contaminated environment. We have given many children self-image tests, and found that, compared to children in less polluted areas, children here have a very limited self image. Their descriptions of themselves – the children here – are not very complex, don't reflect variety and beauty. Of course we cannot prove that this difference is due to the pollution. But we can presume that, in the future, these children will be unable to use their potential for imagination and creativity to its fullest extent. It is very sad and worrying in terms of the future.

At about the same time that Iwonka was telling her story to the class, the Polish minister of the environment, Bronislaw Kaminski, was being interviewed in Warsaw and promising dramatic changes: "First, we must make up for the vast outstanding negligence in expenditure for protection of the environment. We now require that all new industrial complexes built in Poland must be up to current European Community ecological standards." But Poland's present environmental monitoring system,

> apart from giving very alarming results, provides us with only an incomplete picture of the situation. In Silesia, the problem is definitely the worse, if not critical. In some areas of Silesia, agriculture should be limited or even eliminated altogether.

Our research shows that adults are more resilient; those most vulnerable to the ecological threats are children. That is why our new health monitoring system will concentrate first on the children.

SATANIC MILLS

Katowice covers only two per cent of Poland's land, but contains ten per cent of its population. It produces 31 per cent of the nation's coke, 32 per cent of its electricity, 52 per cent of its steel and 98 per cent of its coal, and it mines and processes all of its zinc and lead ores. Heavy metals are breathed in with the air, swallowed with the water and are eaten in the vegetables grown in the gardens everyone plants to supplement the poor supply of food in the shops.

Its people have 15 per cent more circulatory disease, 30 per cent more tumours and 47 per cent more respiratory disease than the average Pole. Oxygen deficiency is the prime cause of death in babies over one month old.[1]

The recent opening up of Eastern Europe has produced a flood of information about the polluted lives there. All East European countries, except East Germany, have seen declines in life expectancy between the mid-1970s and mid-1980s.[2] An estimated one in four Soviet citizens lives in an "environmentally critical" area, and 50 million people breathe air that is ten times more polluted than the nation's own legal limits. Hungary estimates that one in every 17 deaths is related to air pollution. East Germany's newest power-station is thought to pump out more sulphur dioxide than the whole of Sweden.[3]

Life amid these statistics is reminiscent of the horror stories of life around the factories of nineteenth-century industrial Britain. When the health effects of Victorian pollution first came to public attention, the response was that the smoke was a sign of progress and development.

The Victorians also had the same trouble as administrators today in proving a direct link between a given health problem and a given form of pollution. Inadequate monitoring and enforcement made progress slow. Inspectorates were under-staffed and inspectors underpaid; penalties, where they existed, were nominal.[4] It is not hard to find parallels today in East and West Europe and elsewhere in the modern world.

Industrial pollution is increasing rapidly in many Third World cities, due both to old fashioned equipment and lack of enforced emission laws.

It appears that societies must reach a certain level of "development" before they become concerned with the rates of sickness and death caused by pollution. All the major innovations of the past hundred years – the internal combustion engine, nuclear power, the petrochemical industry – were well established *before* their environmental and health effects were even partially understood. Now that they are understood, it would be a massive, preventable tragedy if the South and its young and burgeoning populations were forced to undergo the same dirty industrial revolution that the smaller populations of the North endured.

POLLUTED CHILDREN: THE SCIENCE

Babies and young children, responsible for very little pollution, are much more vulnerable to pollutants than adults, who are responsible for almost all pollution. Most vulnerable of all are the children still in their mothers' wombs.

Infants have a larger surface area in relation to their weight, so they have higher metabolic rates. They process calories and oxygen faster than adults, just as mice eat and breathe more for their size than do elephants. Resting children younger than three inhale about twice as much air as resting adults, per unit of body weight. So they are taking in twice as much air pollution, per unit of weight.

Infants are also vulnerable because their entire bodies and all their internal systems are growing rapidly, particularly during the first six months after birth. Even in fit adults, a high proportion of body weight is fat, which can store many pollutants relatively safely. Young children have very little body fat available, so pollutants circulate longer throughout the body and have greater effects on enzyme systems. Infants are particularly sensitive to toxic chemicals because their kidneys, livers, enzyme systems and blood-brain barriers are not fully developed, and they cannot "process" the pollutants the way adults can.

Children's habits and activities can also increase their exposure to hazards. Hand to mouth contamination is a particular risk in babies and young children, whose instinct is to suck what they pick up. For older children, places of play are often dangerous – the streets because of car fumes and leaded dust, beaches due to sewage polluted sea water, and open spaces due to ill-guarded derelict waste sites and the hazardous substances left there.

Poorer children tend to be most at risk from pollution. They are more exposed because they often live in the most industrialized neighbourhoods, nearest the busiest roads, and – in the Third World – tend to drink the dirtiest water. Poor children are often smaller and thinner, and have even larger surface areas in relation to weight than do "normal" children. And poor children are often malnourished and prone to disease, all of which makes it harder for their bodies to deal with contaminants from outside. Thus, adding air pollution to malnutrition can turn what would be a minor bout of flu in a healthy child into a fatal illness.

The most vulnerable, and innocent, of all are the children still in the womb.

POLLUTION BEFORE BIRTH

Normally, between two and three babies in every hundred are born with some sort of deformity, which may not show up until years later. Of these, about 25 per cent are genetic (passed on from the egg or sperm cells from the mother or father); between five and ten per cent are caused by radiation, viruses, drugs and chemicals; and the rest are the result of unknown causes likely to

Overleaf Malgorzata Swietana with her three children in Poland's highly polluted Katowice region. Krzystof, 8, suffers a slight heart defect; his brother Adam, 11, has a congenital defect that has left him partially paralysed. Malgorzata is worried about the fate of her new baby. *Jacek Petrycki/Central Independent Television PLC*

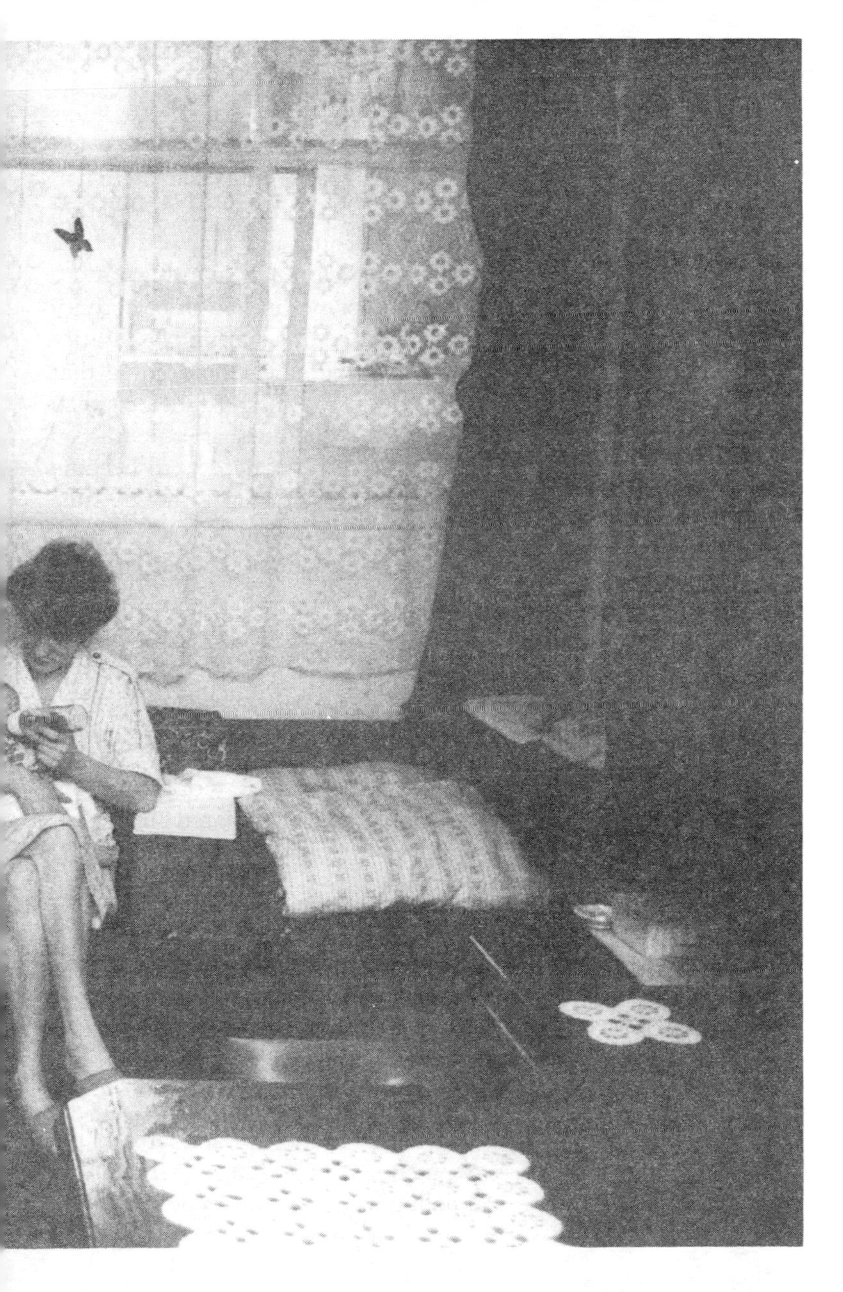

be a combination of environmental and genetic factors.[5]

In recent years some scientists have raised fears that the number of babies born with some physical or mental defects leading to future health problems may be increasing. A US paediatrician published figures in 1983 suggesting that there had been a big increase in the number of children with debilitating illnesses like asthma and chronic bronchitis, and that at that time in the United States there were at least half a million more children than 25 years ago who had some limitation of activity due to either a chronic medical condition or a learning disability.[6]

Many others fail to see any such increase. Some argue that doctors are simply more capable of diagnosing birth defects than previously and, with modern medical techniques, are able to keep more babies alive for long enough for birth defects to become obvious and to be passed on to the next generation. Other possible causes of the apparent increase in birth defects include an increase in smoking among women and the sheer amount and variation of chemicals to which human beings are now exposed.[7]

There is no doubt that each year there are more and more chemicals in the environment which can cause birth defects and damage the reproductive organs of men and women. There is also growing evidence that the toxic effects of chemicals can make themselves felt at very low levels of exposure, in the sort of concentrations that are found in ordinary homes, workplaces and the environment. Heavy metals, polychlorinated biphenyls (PCBs), dioxins, pesticide residues in foods, and other chemical contaminants from a whole range of sources – industry, cars, intensive farming – are inside all of us and throughout our living environments. We are running a large experiment to see what effects emerge from such prolonged low-level doses. We, and especially our children, are the guinea pigs. The overall results of this experiment may not be available for some time yet.

The first threats to children emerge as threats to their mothers' egg cells, or ova, and to their fathers' sperm; both can be damaged by environmental pollution. The risk to the ova and later to the foetus, if the mother is exposed to such things as radiation, tobacco and some chemical pollutants, has been recognised for a long time. Until recently, less attention has been paid to fathers, but the unborn child is also at risk from damage to the father's sperm before conception.

The timing of the risk is different because of the different mechanics of the male and female reproductive systems. Female babies are born with all the egg cells they are ever going to produce already formed. If these cells are damaged before birth, they will still be damaged when they are released as the beginning of a new life decades later. But a woman's healthy egg cells are much less vulnerable than her partner's sperm. Sperm is produced continuously after puberty, so there are endless opportunities for damage to cells as they split and reproduce themselves during this process.

The unborn child is most at risk when it is still an embryo. During the embryonic stage, about the first two months, the different organs of the body begin to form. Each organ system has its own critical time within this period. If the embryo is exposed to a chemical capable of causing a birth defect, the specific defect will depend on which organ is growing most at that time. From two months on, it is the number of cells and the size of the body parts of the developing foetus which are most likely to be affected by exposure to environmental factors. Growth retardation and injury to the central nervous system are the major risks. The brain remains vulnerable throughout, since its development is still incomplete at birth.

The developing child is at risk from two major classes of pollutants, those that cause cancers and those that cause birth defects. Large numbers of chemicals to which mothers are exposed fall into these two categories. Everything the mother inhales or ingests can end up in the blood of the foetus. The umbilical cord carries an artery and a vein linking the foetus's circulatory system to the mother's. A network of tiny blood vessels which run through the placenta and into the womb allow the foetal blood to run so close to the mother's blood that oxygen and nutrients can diffuse across from the mother, and carbon dioxide and waste products can diffuse across from the foetus. The mother's body is capable of insulating the foetus from some toxic substances: for example, fat-soluble substances are likely to be stored in a reservoir of fat cells in the mother, which reduces their concentration in the blood. But these substances are later passed on in increased concentrations in breast milk.

The placenta acts as a barrier against some substances, but many pollutants are found in equal concentrations in the mother and foetus. Recent follow-up studies on children whose mothers who were poisoned with PCBs during pregnancy have shown effects in their children, such as growth retardation, later in life.[8]

Some substances are found in even higher concentrations in the foetal blood than in that of the mother. The mercury levels in the blood of mothers who ate contaminated fish were found to be on average 47 per cent lower than that of their newborn babies.[9]

Some apparently benign substances which do not harm the mother, or cause mild symptoms, may have very serious effects on the foetus. The birth defects found in the babies of mothers who took thalidomide in the early 1960s are a tragic example.

More recently, an increase in the number of spontaneous abortions and of babies with birth defects in California's "Silicon Valley" south of San Francisco in the early 1980s alerted worried parents to the apparent dangers of industrial pollution, even from an industry as clean as the microchip industry. The drinking-water supply in the area was found to be contaminated with chemicals which had leaked from an underground chemical storage tank. After an official investigation, the company responsible paid a large, undisclosed sum to the affected families, though it never admitted liability.[10]

Mothers-to-be and, increasingly, fathers-to-be are told to look after themselves for the sake of their unborn children. But most types of environmental exposure are outside their control. Babies of the most careful parents are now born with their first dose of chemicals already in their bloodstreams, and get their next dose with breast milk. The more scientists learn about the long-term effects of many chemicals, the more they realize that their notions of "tolerance" are often flawed.

THE STRANGE CASE OF LEAD

Lead pollution, and the British response to it, deserves some special attention, not because it is the most intractable pollution problem facing industrialized societies, but precisely because it is not. Technically, it is an easy nut to crack, but many governments, including the British government, have been a long time cracking it, and its history shows where politicians and industrialists tend to set their priorities.

The fact that it has taken so long to get lead out of petrol is an indictment of the ability of governments to protect the health of children in the face of powerful financial vested interests. Only a relatively low value placed on children's well-being could have allowed such complacency. Many of the dangers of acute lead poisoning have been known since reports were published in the late nineteenth century linking lead to high levels of infertility, still births, and spontaneous abortions among pregant women working in lead-using industries.

More recently, the effects of low-level lead exposure have been studied. Dr Herbert Needleman and his associates studied 2,000 Boston school children over 1975–78. He checked the lead content of their teeth, the best indicator of long-term exposure. The children's teachers rated their performance and behaviour in school; parents completed a detailed questionnaire; a battery of tests were given measuring intelligence, academic achievement, listening and language skills, hand and eye co-ordination and attention span. Numerous variables other than lead doses were "levelled" to leave the child's lead content as the key variable. The results were dramatic. In all the tests except one, the low-lead children performed better than the high-lead children.[11]

The British government continued, despite this and other studies, to argue that lead did not pose a significant threat to children's health. In 1980, a working party set up under the chairmanship of Professor Patrick Lawther reported that it had been unable to come to any clear conclusions on the effects of small amounts of lead on the intelligence, behaviour and performance of children. Despite the controversy and criticism which surrounded the report, its findings prolonged the government's complacency.

The petrol industry, meanwhile, continued presenting arguments in its own defence cloaked in scientific dispassion. Five of the largest oil companies – BP, Chevron, Mobil, Shell and Texaco – jointly owned Associated Octel, the company whose works at Ellesmere Port produce much of the world's lead additives to petrol. Octel's chief medical officer, Dr P.S.I. Barry, described by Octel itself as "one of the country's leading authorities on lead and health", said in 1981, well after the Needleman findings:

> Many studies have been carried out to determine whether children are adversely affected by relatively low concentrations of lead in the environment. So far, none of the studies has given positive proof of harm. Some have found no association at all; others have suggested some very slight adverse effects. Any changes noted have always been uniformly small and unrelated to the amount of lead to which the children were exposed.[12]

Two months before this statement, the government's Chief Medical Officer, Sir Henry Yellowlees, had written a confidential letter to senior government officials expressing his concern that, "There is a strong likelihood that lead in petrol is permanently reducing the IQ of many of our children".[13] But for the time being the letter remained confidential, and in March 1981 it was announced, on the strength of the Lawther report, that Britain would not move to unleaded petrol, although the amount of lead in petrol was reduced from 0.4 grams per litre (g/l) to 0.15 g/l.

A year later, the leaking of the Yellowlees letter to the pressure group CLEAR marked a turning point in the campaign. It was published in full in *The Times* in February 1982. This weakened the government's position considerably, but it was not until April 1983 – after the Royal Commission on Environmental Pollution pronounced against lead and with a general election only months away – that the government finally announced a decision to phase lead out of petrol.

THE LEAD TIME LAG

When unleaded petrol first came on sale in Britain in 1986, there was virtually no demand for it. By June 1988, sales still only accounted for about two per cent of the total. It was following the 1989 budget that sales of unleaded petrol really began to increase. A year later, about 30 per cent of all petrol sold in Britain was unleaded. The 1990 budget was the fourth consecutive budget to introduce fiscal incentives to encourage sales of unleaded petrol, which in 1990 sold for about 12 pence per gallon cheaper than leaded.

The government at first resisted the idea of intervention, despite the experience in other countries where fiscal incentives had been the single most influential factor in pushing up sales of unleaded petrol. But pressure from all sides – the environment and industry lobbies – plus the government's desire to be seen to be green during a wave of unprecedented popularity for environmental issues, combined to encourage the government, and particularly the Treasury, to abandon an otherwise resolutely free-market approach.

How could a government possibly take so long to stop "permanently reducing the IQ of many of our children"? The real issue was clear enough. As Needleman put it in 1982: "Action on lead, clearly warranted, long delayed and easily obtained, could be the first in a series of steps to protect the legacy we owe our children – a physical and social world in which their brains and futures can reach the extremity of their potential".[14]

A major reason for the delay was that the government relied heavily for its information on those who had most to lose from getting lead out: the oil and lead-additive industries and the car manufacturers. The oil industry was influential in many government-sponsored studies on the effects of lead in petrol, and the cost of removing it. The technology to produce cars able to run on unleaded petrol had existed for some time. Most British car makers were already producing cars capable of running on unleaded petrol for overseas markets, yet they continued to give the government exaggerated estimates of the time needed to introduce the necesary design modifications.

Overleaf London: more roads, more cars, more pollution . . . and less movement. Balancing the popularity of cars against the environmental destruction they cause poses one of the biggest challenges for democratic governments. *CLEAR*

Hertz
Bruce Grove Tottenham
Stamford Hill Dalston
OLD STREET
YKX 218S

The costs of moving to unleaded petrol were consistently overestimated and the real costs of not taking lead out of petrol simply did not figure on the balance sheet. These costs are difficult to add up, but other governments had at least tried. In the United States in the early 1980s the Environmental Protection Agency (EPA) came under pressure from industry to justify its earlier decision to regulate a lead phase-out. During the row, a leaked memo revealed that the EPA's Office of Policy and Resource Management had estimated that allowing lead to stay in petrol at then current levels would save industry $100 million a year. But the cost of treating the additional 200,000 to 500,000 children that would be poisoned by lead would be between $140 million and $1.4 billion.

GETTING THE REST OF THE LEAD OUT OF BRITAIN

The lead battle is not over even in Britain. Today leaded petrol is being phased out, and lead has largely been removed from paint, so that inquisitive children no longer suck on brightly coloured chips of lead paint except in the oldest buildings. But old lead pipes are still delivering a steady supply of leaded drinking water to 45 per cent of the 18 million households in England, Wales and Scotland. Some 2.5 million of these homes are in soft water areas and are particularly at risk, since soft or acidic water dissolves lead more effectively than hard, alkaline water. Nationally, water supplies about six per cent of the lead humans consume. But in soft water areas that figure rises to more than 50 per cent of total body lead.[15] The UN describes lead plumbing in soft-water areas as a "major potential source of exposure to bottle-fed infants".[16]

Britain's post-privatization water regulations set a maximum level for lead in drinking water of 50 micrograms per litre. This is equivalent to the current World Health Organization (WHO) standard, and less than the European Community limit of 100 micrograms per litre. In the last national survey of domestic tap-water in England and Wales in 1975–6, 7.8 per cent of samples exceeded the 50 microgram limit and 2.6 per cent of samples exceeded the 100 microgram limit. In Scotland, where soft water makes lead in water a particular problem, 34.4 per cent of samples exceed the 50 microgram limit and a further 26 per cent exceed the 100 microgram limit.[17]

In the most recent major study of the relationship between lead and health in Scotland, water was found to be putting more lead into people than was petrol. Eight hundred Edinburgh schoolchildren aged between six and nine were tested. Children with higher blood lead levels performed less well intellectually than those with lower levels, and scored less well on number skills and reading ability. A later stage of this study also showed a clear link between blood lead levels and aggressive, anti-social and hyperactive behaviour in children.

These effects – including a difference of about six IQ points between high and low lead levels – occurred at what were previously considered very low levels of lead in blood. There was no indication of a threshold below which effects would cease.[18]

In 1988 the US Environmental Protection Agency recommended an average lead-in-water limit nearly five times lower than Britain's maximum acceptable concentration of 50 micrograms per litre.[19] In Britain, a 1989 working party report to the Ministry of Agriculture recommended that the *average* lead-in-water limit for bottle-fed babies should be 10 micrograms per litre, and for the population at large, 30 micrograms per litre.[20] There is no doubt that some areas of Britain, particularly soft-water areas, have average lead-in-water levels that exceed these recommendations.

In 1983, the Royal Commission on Environmental Pollution made some tough recommendations on lead-contaminated water. These included early completion of surveys to identify areas with high lead levels in drinking water; early and firm target dates for completion of a remedial programme of water treatment or pipe replacement (acidic waters can be chemically treated before they enter the mains system, to reduce their capacity to dissolve lead); review of criteria for home improvement and repair grants to make grants more widely available; and publicity campaigns for areas where pipe replacement is needed. The Royal Commission said that financial constraints should not be allowed to hamper the pipe replacement programme.

Little has been done to implement these regulations. Grants are still discretionary and limited. Publicity about the hazards of lead in water is virtually non-existent. Remedial measures have concentrated on water treatment rather than pipe replacement. No extra funds have been given to local authorities to carry out remedial work. No surveys of schools, hospitals or residential homes have been published.

The new UK upper limit of 50 micrograms cannot be met without an expensive pipe replacement programme. Currently the burden is left largely on the householders. The new regulations specifically exempt the water supplier from any liability with regard to domestic plumbing. Only if the householder decides on lead pipe replacement must the supplier replace pipes servicing the property. In 1981 The National Water Council estimated the cost at £600 per dwelling, of which two-thirds was the householder's responsibility. At today's prices if all remaining "at risk" households took up this option, the water suppliers' bill could run into hundreds of millions.[21]

The government has accepted undertakings from the newly privatized water companies which give the timetable for compliance with the lead standard. Water companies have promised to identify zones where there is a problem by 1991. But no date has yet been set by which all supplies must meet the lead standard.[22]

LEAD IN THE WIDER WORLD

Lead pollution, a threat primarily to children, remains a problem worldwide. Eight out of 23 cities monitored by the United Nations between 1980 and 1984 were on the borderline or over the WHO limit. Some European countries – France, Italy, Spain, Belgium and Portugal – have been even slower than Britain in bringing in unleaded petrol, despite pressure from a European Community Directive (EC Directive 85/210/EEC). The UN study found airborne lead levels in Paris more than five times higher than the WHO limit.[23]

Lead pollution is increasing rapidly throughout Latin America as car and truck traffic in many cities approaches European levels; it is also increasing in many African cities. Much of the petrol in these countries still contains up to 0.84 g/l of lead, a level that has not been seen in Britain since 1972. Air and blood lead levels in Third World cities are rising, compared to falling levels in much of the North.

People living in about one-third of the cities of the world are likely to be exposed to lead levels in the air which are either marginal or unacceptable, WHO estimates.[24] Seven out of ten newborns in Mexico City have blood lead levels higher than WHO norms.[25] And petrol in Mexico City had the highest lead content of all sites tested in the WHO survey.

TRANSPORT: GETTING FROM HERE TO THERE

Lead is only one pollutant from the internal combustion engine. It is the easiest to deal with; it is taken out of the fuel. To produce unleaded petrol, the oil and motor industries had to change the way they made petrol and cars. Changing the way industrial society works – that is, encouraging millions of people to drive less in their cars, and to make them pay more for the privilege of doing so – will be infinitely more difficult. As Ian Breach wrote:

> The car makes more demands and inflicts more damage on our global habitat than any other commodity – but it is universally the most desired artifice in the whole history of humanity. Its grip on our culture, our economy, and our environment is comprehensive – and comprehensively daunting.[26]

There is no complete technical cure for the pollutants produced by the internal combustion engine: nitrogen oxides, the hydrocarbons that help to make smog, the particles, and the planet-warming carbon dioxide. Catalytic converters, lean-burn engines, battery-driven cars, gas turbines, alternative fuels and other technologies may all help. But the only long-term solution short of solar-powered or hydrogen-powered engines is to change the mix of transport so that there are fewer cars and more – and more efficient – forms of public transport, and perhaps more – and more frequently used – bicycles.

Future global transport development cannot be based on the car without disastrous results. The 400 million cars on the world's roads today are already putting an intolerable burden on the atmosphere, both local and global. If Indians owned cars at the same people-to-car ratio as US citizens (two people to every one car), then Indian roads by themselves would be bumper-to-bumper with well over 400 million cars. The Chinese would be driving more than 500 million.

For the sake of the planet, every young Indian cannot be allowed to realize the dream of every young American or European: private ownership of a petrol-burning car. But how can we preach such realities while continuing to encourage car ownership in Northern countries which ought to know better? Speaking at the "Better Environment Awards for Industry" in March 1990, Mrs Thatcher confirmed there were no plans to "do away with the great car economy".[27]

The United States, Canada, Europe and Japan account for 16 per cent of the world's population but produce 88 per cent of its cars and own 81 per cent of them. Every day more than 100,000 cars roll off the world's assembly lines, accounting for more than 12 per cent of world trade in manufactured goods.[28] About one per cent of people in the poorer countries of the South own cars, compared to 40 per cent in the North. But since 1979, the most rapid growth in car ownership has been in Asia and South America, where it has more than doubled.[29]

Cars are an inefficient means of getting around. Only 10–20 per cent of the fuel put into them is converted to energy; the rest comes out as pollution, heat and noise. A big car (with an engine over 2000 cc) with one person in it uses six times as much energy per passenger mile as a one-third full double-decker bus. Even a two-thirds full Boeing 737 uses less energy per mile, per person, than a big car carrying only its driver.[30] With these facts in mind, the industrialized North has chosen the car as the basis of its transport policies.

"The motorist is an individual, and he likes being a motorist because he can exercise his freedom of choice", said Nicholas Ridley when he was British Transport Minister. "That is a good thing. He should not be hampered by petty rules and restrictions on his liberty."

But at what cost "his liberty"? Every time someone chooses to drive a car, public transport delays are aggravated by increased congestion on the road; more money is diverted away from public transport, leading to fewer and poorer services; and everyone, drivers and non-drivers, breathes in more pollution.

Freedom of choice even for motorists is becoming meaningless as it becomes difficult to move on the roads of major cities. In London, traffic moves at an average crawl of about 10 miles per hour.[31] Underfunded buses, trains and underground systems are also inefficient, uncomfortable and overpriced. The choice for car drivers is between slow-moving frustration behind the wheel of a car or poor public transport. For people without cars, who include all unaccompanied children and many children in the care of women, there is no choice. The sole option is inadequate public transport.

Expenditure on the national roads system in Britain increased by 36 per cent between 1982/3 and 1988/9 to reach £1.3 billion. In the same period, government spending on local roads also increased, but spending on local public transport dropped from £614 million to £347 million. Levels of investment and the ways in which people choose to travel tally closely. The use of both public transport and bicycles in Britain has declined since 1960, while the use of private cars has more than trebled. In 1988 journeys by train accounted for only seven per cent of all travel in Britain.[32]

The British tax system actually subsidizes the car, the transport of the wealthiest. In 1989 company cars were under-taxed to the tune of approximately £2 billion, some four times the amount the taxpayer gives British Rail in public subsidy. This worked out at a subsidy of about £800 of taxpayers' money for every ton of carbon dioxide produced by a company car.[33]

If an employer provides a company car, neither the company nor the employee pays National Insurance contributions on its value. In addition, the cash equivalent or "scale charge" set against it for income tax purposes is much lower than the true value of having the car. Free parking at the place of work, a valuable perk, is entirely free from tax and national insurance.[34]

In April 1990, the scale charges for company cars were increased for the third year running, this time by about 20 per cent, reducing the £2 billion in lost revenue on company cars by about £160 million. Although welcoming this as a step in the right direction, environmental groups were disappointed that more substantial changes were not introduced.

In 1988 there were 2–4 million cars registered with companies, 30 per cent more than in 1983.[35] Sixty per cent of all new car sales in 1988 were also to companies.[36] Company cars tend to be larger than privately bought cars, to be driven faster, to have more accidents and to be used more extravagantly.[37]

The incentive to drive to work and park in city centres causes distortions in transport patterns and adds to pressure for more roads. A survey of central London car users in 1982 found that 79 per cent of those commuting by car were receiving some form of subsidy from their employers towards the cost of motoring. If cars owned by the self-employed are included in the calculation, more than one in three cars on the road are effectively government-subsidized.[38]

The inequity is rooted in the way in which transport policy decisions are made. Major roads are justified on the basis of calculations of how much faster people will be able to travel, and hence how much time and money will be "saved". But different classes of road user are valued at different levels. Car drivers' time is worth more than that of bus passengers. Until 1987, pedestrians and cyclists were not given any value at all. A pensioner's time is assessed at three-quarters the rate of someone working, and a child is worth only a quarter of this rate.[39] The inequity is greatest for those least able to afford it. In Britain in 1987, about two-thirds of households had a car and more than a third of these had two or three cars; 36 per cent of households rely on public transport.[40] The freedom enjoyed by those with cars has become a burden for those without.

Abolishing tax relief on company cars, introducing fiscal incentives to leave cars behind and to drive cleaner cars, and the implementation of "carbon taxes" would all help to get more people out of cars and on to public transport. Other countries have experimented with road pricing systems – charging on the basis of pay-as-you-drive. When road pricing was introduced in Singapore in 1975, rush-hour traffic going into the city fell by 44 per cent. Many commuters took to car pooling, and bus travel rose sharply. The system is now firmly established.[41]

It is hard to imagine the British government encouraging cycling only because the British government never has. Other governments running prosperous, modern nations do. Such policies are not so much anti-car as pro-bicycle, as in Denmark, The Netherlands and West Germany. In The Netherlands and Denmark, up to 20–30 per cent of all urban trips are now made by bicycle – up to 50 per cent in some towns.

Increased use of bicycles would cut down particularly on short car journeys within cities, precisely those that create the most pollution because of low speeds and cold engines. A move to bikes for these journeys would improve city air disproportionately. Commuting by bike has massive potential in industrial cities: more than half of all commuter trips in the US and nearly three-quarters of those in Britain are less than five miles.[42]

TRAFFIC – THE OFFICIAL RESPONSE

The British government has forecast further massive increases in car numbers – between 83 and 142 per cent by the year 2025 – and called for more roads.[43]

In early 1990, the Department of Transport confirmed its plans to spend £12.4 billion on a road-building programme for the next decade. The government line is that congested roads cause more pollution, so they must build more roads to ease congestion. But transport economists are generally agreed that the demand for car journeys expands to meet the road space available. In Britain research has shown that even with three-way catalytic converters fitted to new cars, after an initial decrease, total emissions of nitrogen oxides will start rising again early in the next century if the number of cars on the roads increases according to government expectations. Carbon dioxide emissions cannot be reduced by converters. Many argue that more and wider roads will simply mean more miles driven at the expense of other less polluting forms of transport.

In 1986, 407 children were killed and 41,510 children injured on British roads. This represented a quarter of all deaths to children between 5 and 14 that year. Despite more and faster cars, the number of children killed and injured on the roads has been substantially reduced in recent years, but a large part of that reduction has been achieved at the expense of children who can no longer play in the streets, or visit friends or go to a park without an adult. This is particularly restrictive in residential areas, where most accidents involving children happen.[44] The rate of casualties to children is highest around 8.00 a.m. and between 3.00 and 5.00 p.m. as they travel to and from school.[45]

Not surprisingly the Department of Transport's enthusiastic road-building programme has brought it into conflict with the Department of the Environment. Such a dilemma is endemic to systems of government which divide human needs into strictly compartmentalized ministries and departments. The World Commission on Environment and Development summed up the "ministry syndrome" in speaking of all environmental challenges: "... most of the institutions facing these challenges tend to be independent, fragmented, working to relatively narrow mandates with closed decision processes".[46] British Environment Minister Chris Patten admitted that he suffered the same syndrome when he said, somewhat wistfully: "As Secretary of State for the Environment, I am acutely aware that some of our biggest problems are handled elsewhere".[47]

In the meantime car-centred development and the environmental stress that accompanies it continues virtually unchallenged and unreconciled to the need for protection of the environment.

In developing countries, road traffic deaths and injuries to children are increasing, in line with growing car use. The number of accidents in Korea, Turkey, Saudi Arabia, Togo and Senegal increased by more than 150 per cent between 1978 and 1985.[48]

AIR POLLUTION IN THE 1990s

In the United States, some 150 million people breathe air considered unhealthy by the US EPA, according to a world-wide pollution survey by Hilary French.

In greater Athens, the number of deaths is six times higher on heavily polluted days than on those when the air is relatively clear. One in twenty-four disabilities in Hungary is caused by air pollution, according to government estimates. In India, breathing the air in Bombay is equivalent to smoking ten cigarettes a day, and smoking ten cigarettes a day is much more harmful for babies than for adults. Mexico City is considered a hardship post for diplomats because of its unhealthy air, and some governments advise women not to plan on having children while posted there.[49]

In Britain, over ten million people – one in five of the population – are at risk from polluted air, according to a 1989 Friends of the Earth study. The most vulnerable groups include children under two years old, the elderly, pregnant women and the babies they carry, and people suffering from illnesses such as asthma, bronchitis, emphysema and angina.[50]

The link between respiratory diseases in children and local air pollution has been well documented. "There is by now much evidence of an association between the general prevalence of respiratory illnesses in children and the amounts of pollution in the areas in which they live", said a 1974 paper from the Medical Research Council's Air Pollution Unit. Contracting respiratory diseases in childhood can lead to lifelong problems.[51]

Most harmful things inhaled today come from industrial processes, power plants or vehicles. Some air pollutants are produced by all three.

TRANSPORT POISONS

In addition to lead, vehicles also emit carbon monoxide, nitrogen oxides, hydrocarbons and carbon dioxide. Nitrogen oxides and hydrocarbons are toxic in themselves, but they also react together in sunshine to form ground-level ozone, one of the main ingredients of smog. (The word "smog" is often preceded by the word "photochemical" because sunlight powers the chemical reactions which produce it.)

Scientists are trying to find ways to get rid of ground-level ozone, which is hazardous to health, and to preserve ozone high in the stratosphere, which protects living things from too much ultra-violet radiation. Ozone has become a major health problem in and downwind of big cities worldwide. At high concentrations, it can cause severe lung damage and make bacterial and viral infections more likely. At lower concentrations, it irritates the lining of the respiratory system, causing coughing, choking, impaired lung function, irritation of the eye, nose and throat, and headaches, particularly in people who exercise. It also makes asthma and bronchitis worse.

Recent US studies have shown that health effects occur at lower levels than previously thought. The lung functions of children and asthmatics particularly can be affected at relatively low concentrations. And long-term study of rats suggests that ozone effects are cumulative – that is, each exposure adds to the effects of previous exposures.[52]

Ozone concentrations are increasing North America and Europe, and recent hot summers have seen record levels.

Information on ozone in Eastern Europe, USSR and the developing world is scarce, but conditions are bound to get worse in the immediate future as the number of vehicles on the roads increases. In much of Latin America, a combination of climate and recent increases in the number of cars on the road means ozone is already a serious problem.

NITROGEN OXIDES

As well as contributing to ozone, nitrogen oxides have similar direct health effects: increased risks of viral infections, bronchitis and pneumonia, as well as general lung irritation.

There has been little long-term monitoring for nitrogen oxides in Britain, but concentrations appear to be increasing, particularly in London. Annual average concentrations rose by a quarter between 1980 and 1985. Concentrations of nitrogen oxides are likely to be above the WHO guidelines at busy roadside sites in London, according to the Warren Springs Laboratory (WSL), the government's monitoring agency.[53]

In Britain, cars are responsible for 45 per cent of nitrogen oxides, with another 35 per cent coming from power stations. Even though emissions from power stations have dropped over the past ten years, this has been more than offset by more pollution from cars.[54] Because cars pollute at ground level, emissions from cars have a greater local impact.

Repeated short bursts of high-level exposure to nitrogen oxides are more dangerous to health than the total long-term exposure at lower levels. Some 15–20 per cent of North Americans and Europeans are exposed to unhealthy levels of nitrogen oxides, according to UN estimates.[55]

CARBON MONOXIDE

The effects of carbon monoxide (CO) are known to anyone who has been stuck in a car in heavy traffic for even a short time: drowsiness, blunted perception and thinking, slowed reflexes and headaches.

The blood takes up CO in preference to oxygen, so at very high levels, CO poisoning can be fatal. Babies, children and the unborn children of pregnant women are among those most at risk, as well as the elderly, and people suffering from heart or respiratory diseases. CO inhaled by pregnant women passes in the blood to the unborn child and is thought to retard growth and mental development.

At least 90 per cent of CO worldwide is produced by vehicles, and in some cities virtually 100 per cent.[56] Levels peak during morning and evening rush hours, and higher levels are often found inside cars, where the air is trapped, than outside in moving air. In Britain CO emissions increased by about ten per cent between 1977 and 1987 because of increased traffic.[57]

Even in countries where catalytic converters are required on cars, CO remains a problem. Half the people living in cities in North America and Europe are exposed to unhealthy levels of carbon monoxide, according to UN estimates.[58]

DIESEL FUMES

The other main source of pollution from traffic is particulate emissions or smoke – carbon particles less than one micrometre wide – pumped out from diesel vehicles. Diesel-burning heavy lorries, buses, taxis and cars emit more than ten times more of these particles than a petrol-driven car and over 100 times more than a car driven on unleaded petrol.

Inhaled particulates can cause breathing problems. All particulate pollution can also carry other toxic subtances into the lungs, so exposure to diesel fumes has been implicated in cancers, though industry and power stations are the major overall sources of particulate pollution.[59]

In Western Europe and the United States, smoke emissions from diesel engines are rising. In some cities, diesel vehicles account for more than 70 per cent of all smoke pollution, with industry providing the rest. The United States regulates the chemical content of diesel emissions, whereas less effective European regulations focus only on the density of the smoke.[60] Diesel pollution is a major problem in many Third World cities, as badly maintained engines pollute more than those that are well cared for. Fleets of buses and trucks in the Third World tend to be old and badly maintained.

EFFICIENT CATS

A number of technologies have been and are being developed to reduce the amount of pollution produced by vehicles. Of the present clean-up measures, catalytic converters, popularly known as "cats", are the most significant.

A cat is fitted in an exhaust pipe. It consists of a metallic or ceramic honeycomb coated with a very thin layer of precious metals; it has a surface area the equivalent of two football pitches. As the pollutants pass through the honeycomb, contact with the precious metals causes a series of chemical reactions. CO becomes carbon dioxide; hydrocarbons become carbon dioxide and water; and nitrogen oxides become nitrogen. Cars fitted with cats must run on unleaded petrol, as the lead would spoil the precious metals. The most sophisticated of them, the regulated three-way cat, can reduce pollution – except carbon dioxide – by about 90 per cent.

In 1989, after five years of heated debate within Europe, the European Commission finally agreed that standards for all new cars would be tightened to US levels from 1992 onwards. To meet these standards, cars will have to be fitted with regulated three-way catalytic converters.

POWER STATION POLLUTION

Sulphur dioxide and airborne particulate pollution are the two main industrial pollutants worldwide. Ninety per cent of sulphur dioxide pollution comes from power stations burning fossil fuels; the vast majority is produced in the industrial countries. Sulphur dioxide and particulates are often present together in industrial cities, a combination which aggravates the health effects of both. So most studies consider their effects together.

The direct health effects of inhaling sulphur dioxide include chest and respiratory diseases such as bronchitis. Asthmatics suffer more than most, and long-term exposure can raise death rates from heart and respiratory diseases. WHO is particularly worried about children continually exposed to concentrations of sulphur dioxide above its guidelines, as early respiratory diseases caused by air pollution are thought to contribute to chronic respiratory disease later in life.[61]

Fine sulphate particles formed from sulphur dioxide liquefy and become aerosols that can penetrate the lungs, taking toxic metals and gases with them. Some researchers estimate that this toxic mix may be responsible for up to 50,000 deaths in the United States every year.[62]

A combination of sulphur dioxide and smoke, produced by burning coal and oil, caused the thick pea-souper smogs that polluted London and other British cities for more than 100 years. It was only after the infamous London smog of 1952, which left 4,000 people dead and tens of thousands of people ill, that the government took action. In 1956 the Clean Air Act was introduced. Smoke emissions have since dropped by about 85 per cent and sulphur dioxide emissions by about 40 per cent.

Changes in fuel use, lower demand for electricity for industrial purposes, and energy conservation have all contributed to this. Other Western cities have shown a similar trend. Between 1970 and 1985, the United States and West Germany both reduced sulphur dioxide emissions by 27 per cent. Japan reduced its emissions by 14 per cent just in the five years between 1980 and 1985.[63]

Most countries built tall stacks to improve local air quality, rather than investing in technology to reduce emissions; so they simply sent their health problems elsewhere. Although acid rain, caused by sulphur dioxide and nitrogen oxide pollution, is best known for its effects on trees and vegetation, it is now suspected of having serious health effects also. The acidic rain dissolves several dangerous metals – including aluminium, cadmium, mercury and lead – and carries them from the soil into groundwater, streams and reservoirs, potentially contaminating water supplies and killing fish and other water life.

Reductions in sulphur dioxide emissons must be seen in the context of continuing problems. In London, airborne levels regularly exceed the one-hour WHO guideline of 350 micrograms per cubic metre. The highest level measured in 1989 was almost double the WHO limit, at 672 micrograms per cubic metre. Other European cities experience high peak concentrations of sulphur dioxide; Berlin in 1987 suffered peak half-hour levels of up to 1,400 micrograms.[64]

Twenty-seven of the 54 cities with sulphur dioxide data available for 1980–84 exceeded or were on the borderline of the WHO daily average standard of 150 micrograms per cubic metre, not to be exceeded more than seven days a year, according to UN figures in 1988. Seoul, Tehran, Shanghai and Shenyang were all among the worst offenders; but European offenders included Paris, Milan and Madrid.[65] Some 625 million people around the world are exposed to unhealthy levels of sulphur dioxide, with many more on the margins.[66]

The ten cities with the highest levels of particulate pollution in the UN survey are all in Asian developing countries, while nine out of ten of the cleanest cities are in developed countries. Average annual concentrations were as much as five times the WHO standard in Kuwait, Shenyang, Xian, New Delhi and Beijing. The extraordinary levels measured in some Third World cities can be partially explained by natural dust, but diesel-fuelled vehicles without even basic pollution controls, and scooters with highly polluting two-stroke engines, are also to blame. The UN estimates that more than a billion people – one fifth of the world population – are exposed to raised levels of particulate pollution.[67]

Conditions in some Third World cities continue to deteriorate. In India, sulphur dioxide emissions from coal and oil have nearly tripled since the early 1960s. In most of the Third World, pollution control regulations, and the enforcement of any regulations which do exist, are minimal. The Third World is moving into cities, many of which are rapidly industrializing; 90 per cent of future population growth will occur in the Third World. All of which means that in the developing world, pregnant women, newborns and infants in their millions will suffer gradually worsening pollution for the foreseeable future.

One of the best ways to reduce sulphur dioxide from power-stations is to fit flue-gas desulphurization technology, popularly known as "scrubbers", on coal-burning power plants. Scrubbers can remove up to 95 per cent of a given plant's output of sulphur dioxide. A variety of other technologies for controlling sulphur dioxide and nitrogen oxide emissions from power plants is being investigated. Controls of particulate emissions from power plants are now required in virtually all OECD countries.

It is easier, and cheaper, to build new plants with anti-pollution technology than to adapt old plants. But "retrofitting" old plants is important since the building of new plants has recently slowed down considerably. At the beginning of 1987 the percentage of electricity from coal-burning plants produced by plants fitted with scrubbers was about 40 per cent in West Germany, 50 per cent in Sweden, 60 per cent in Austria and 85 per cent in Japan, but only 20 per cent in the US and none in Britain.[68]

In 1988, after five years of difficult negotiations, a plan for reducing sulphur dioxide emissions by fitting power stations with FGD technology was agreed within the European Community. Reductions were planned in three stages, by 1993, 1998 and 2003, but different countries were allowed to reduce their emissions by different amounts, depending on circumstances and fuel supply, as some types naturally produce less sulphur than others, making it easier to meet lower pollution levels.

Air pollution from fossil-fuel burning is spectacularly high in many cities in Eastern Europe and the Soviet Union. Rapid industrialization after the Second World War and a reliance on coal (especially brown coal, or lignite) are both factors.

The rapid oil price hikes in the 1970s forced Western industries to invest heavily in energy efficiency. But in Eastern Europe, oil prices were held down artificially by the state, so that energy remained cheap and there was no market encouragement to use less of it. For example, half of Eastern Europe's steel production still relies on the open-hearth furnace, an energy-wasteful technique now virtually abandoned in the West.

With the collapse of the Communist regimes, market forces are already having an effect. In January 1990, Poland raised its energy prices almost to world market levels; industry is already investing in energy savings.[69]

INDOOR AIR POLLUTION

The effects of indoor air pollution, especially on women and children in the Third World, may be worse than outdoor pollution. Where traditional stoves fuelled with wood and dried animal dung are used for cooking and heating, with little ventilation, particulate pollution and hydrocarbons released as smoke are the biggest problem, along with carbon monoxide, nitrogen oxides and formaldehyde. The WHO guideline for indoor particulate pollution is 100–150 micrograms per cubic metre; many rural houses in developing countries have levels of between 300 and 14,000 micrograms. Chronic lung disease, heart disease, cancer and acute respiratory infections, particularly in children, are all increased significantly by indoor air pollution.[70]

WHO estimates that at least 400–500 million people worldwide are affected by these problems. Women and children are more at risk because of the time they spend indoors and near stoves. Health reports from Burkina Faso and Zambia, among others, confirm that respiratory illnesses are among the top two or three complaints in children. For pregnant women, air pollution is one of the factors that can contribute to lower birthweight babies, who will then be more likely to suffer from a whole range of infant illnesses; the risk of birth defects may also be increased.[71]

TOXIC WASTE – DRINKING WATER AND PLAYGROUNDS

Not all pollutants come out of the air; many of the toxic substances that children are particularly vulnerable to end up in food and water or are picked up on contaminated land.

Children are exposed when they play on sewage-contaminated beaches, or derelict factory sites and waste tips which conceal hazardous substances. Even children's playgrounds can be risky. More than 100 people, mainly children, go partially or totally blind every year in Britain after ingesting the eggs of a parasite (*Toxocara canis*) found in dog faeces. These eggs can survive on the ground for months and easily pass from hand to mouth when children are playing in parks. Another type of toxocara infection causes wheezing and skin rashes.[72]

Virtually all waste – industrial, agricultural or domestic – is potentially hazardous if it is not treated and disposed of correctly. In Britain the most toxic of all toxic waste, "special waste", is defined by the government in terms of how much would kill a child: "a single dose of not more than five cubic centimetres would be likely to cause death or serious tissue damage to tissues if ingested by a child of 20 kilograms body weight".[73] Most of this kind of waste is incinerated, a process which has its own health risks.

In Britain, most less toxic but still hazardous waste is dumped in landfill sites or tips, often in breach of safety regulations, and can contaminate land and pollute drinking water. A 1990 report from the House of Commons Environment Committee estimated that up to 50,000 hectares of land in Britain may be contaminated with toxic wastes.[74]

Children are at risk particularly because of their greater susceptibility to polluted water and because contaminated land is unregistered and often unprotected. There have been many cases of children playing on badly managed, leaking waste tips.

Britain's lax attitude to burying old waste has been under fire from environmentalists for years. The liquids that seep out of buried industrial waste and decayed domestic waste can cause serious water pollution problems. Highly toxic materials can be a danger even years after a tip has closed; if groundwater is contaminated the damage is often irreversible. "All water authorities report groundwater pollution problems with landfill sites to a varying degree, and in many cases it is regarded as the most significant threat", said a 1988 government report.[75]

Hundreds of thousands of British children drink tap water which contains pollutants above European Community safety limits. The main problems are caused by polluted water supplies from intensive farming and leaking toxic waste tips, and contamination which occurs in the water distribution system, as for example from lead piping.

The EC Directive (80/778/EEC) on drinking water quality sets maximum admissible concentrations (MACs) for 66 features of water. As well as lead and nitrates (see below), these include bacteria, aluminium (which has been linked to Alzheimer's disease), cadmium, pesticides and certain suspected carcinogenic chemicals. Babies are more vulnerable to any health risks from contaminated water and drink much more water than adults do in relation to their body weight.

The drinking water directive was issued in 1980 to allow standards to be met by 1985, but maximum levels were still being frequently breached in 1990. The European Commission has protested at the British government's failure to meet standards, particularly for nitrate and lead levels in the most seriously affected areas of the country.

NITRATE: RED BLOOD AND BLUE BABIES

As farmers use more and more nitrate fertilizer and manure, more and more nitrate gets into drinking water. In Britain, use of nitrogen fertilizer increased from 60,000 tonnes a year in the 1940s to a 1985 total of about 1.58 million tonnes.[76]

About half the nitrate fertilizer added to the soil is taken up by growing plants. The rest either leaches into groundwater deposits far below the ground's surface or is carried off fields by rainwater and into streams and reservoirs. In either case, much of it reappears in drinking water.

Nitrate itself is not particularly toxic, but bacteria in the mouth or stomach convert it to nitrite. This combines with the red pigment haemoglobin, which carries oxygen in the blood, to form methaemoglobin. The blood thus carries less oxygen, which can lead to a potentially deadly condition known as methaemoglobinaemia, or "blue baby syndrome". Babies under three months old are particularly at risk because the haemoglobin in their blood combines more readily with nitrite and because they have lower levels of a protective enzyme which reverses the effect of nitrite on red blood cells.

Most recorded cases of blue baby syndrome have been in Third World countries where well-water containing high concentrations of nitrates was used to make up baby-milk substitutes. In most such cases, water containing more than 90 milligrams per litre (mg/l) of nitrate was used, although there have been some instances of the syndrome where milk powders were mixed in water containing less than 50 mg/l. The EC limit is 50 mg/l, but much British water exceeds this limit. WHO has recommended that dried baby milks should be mixed with low-nitrate water (at least below 45 mg/l) and that low-nitrate vegetables should be used in baby foods.

Blue baby syndrome is very rare in Britain, but there is concern that newborn babies could suffer a degree of oxygen deficiency without showing any clinical symptoms. Studies have found methaemoglobin in the blood of children exposed to drinking-water nitrate levels of 44–88 mg/l, yet no proper studies of sub-clinical problems have been carried out. Older children and pregnant women may also be at risk from the formation of methaemoglobin.

In 1988, a confidential Department of Health report leaked to Friends of the Earth warned of cases of blue-baby syndrome where nitrate levels were below the official hazard level. It advised health officials to "remain wary about water supplies containing nitrate in the 50–100 mg/l range".[77] In 1988, villages in Yorkshire had to be supplied with bottled water for babies under six months old because of the high nitrate levels in local supplies.

In East Anglia, babies have been prescribed bottled water, and the local water authority has two mobile bottling plants ready to supply babies if nitrate levels increase.[78]

More than 900,000 people in Britain, mainly in the South-east, were regularly receiving drinking water which exceeded the EC limit of 50 mg/l in 1988, according to the government. But many more people – 1.6 million in 1987 – were occasionally supplied with water in breach of the tighter WHO guideline of 44.5 mg/l.

The situation is likely to get worse; nitrate levels in rivers in the most seriously affected areas – mainly East Anglia and the Midlands – are still rising. The number of groundwater sources severely polluted by nitrate more than doubled in Britain between 1976 and 1986.[79]

Nitrate enters groundwater after dissolving in rainwater and slowly percolating down through unbroken rock. This process can take many years. Today's levels of nitrate in groundwater therefore often reflect the amount of nitrate fertilizers being used up to 20 years ago. Since then, fertilizer use has increased by about 800 per cent.[80]

PESTICIDES AND CHILDREN

The use of modern pesticides has increased steadily over the past two generations, leading to steady increases in the daily human consumption of pesticides. There are many reasons why the results of this will be seen first in children.

The modern pesticide revolution began after the Second World War. Since then, farmers have got stuck on an ever-accelerating pesticide treadmill: the more they use, the more resistant the target pests become, and the more they must use. Farmers are not required to label fruit and vegetables that have been treated with pesticides.

Pesticides are, by definition and by the nature of their job, poisons, and most are toxic to people. Like any poison, pesticides in high doses can have acute, immediate effects including nausea, giddiness, or even unconsciousness. But the longer-term, low-level effects are causing the most anxiety. Specific fears include the threat of birth and genetic defects, increased risks of cancer, allergies, psychological disturbance, and possible damage to the body's defences against diseases.

Ninety per cent, by weight, of fungicides used in America have some tumour-forming capacity, according to a 1987 study by the US National Academy of Sciences.[81] Exposure to pesticide residues may have chronic effects years after the original exposure.

Although many pesticides are used in combination, the possible effects of these chemical cocktails on human health has hardly begun to be examined. It is a recognized premise of science that chemicals in combination can produce effects greater than the sum of the effects of the individual chemicals, and also different from those effects. Many countries, including Britain, continue to use pesticides which have been banned elsewhere. Even when a pesticide is banned in any one country, it may still be inadvertently imported as residues in fruit and vegetables grown abroad.

In the late 1980s, the US Natural Resources Defense Council (NRDC), an independent pressure group, spent two years evaluating the risks to health of the pesticides contained in 27 types of fruit and vegetables eaten by children under five.[82] The council found that the average child receives four times more exposure than an adult to eight widely-used pesticides suspected of causing cancers. Because of their exposure to these pesticides alone, up to 6,200 children may develop cancer some time in their lives. These eight pesticides are just a fraction of the 66 that the US EPA has identified as potentially cancer-causing and which might be found in a child's diet.

The greatest source of cancer risk identified by the NRDC survey came from apples, apple products and other foods such as peanut butter and processed cherries that may be contaminated with daminozide and UDMH, the substance daminozide breaks down into during processing. The trade name for daminozide is Alar. As a result of the NRDC study, Alar became the focus of consumer campaigns in the US and Britain by such groups as Mothers & Others for Pesticide Limits and Parents for Safe Food. The pesticide's manufacturers, Uniroyal, announced a voluntary ban on sales of Alar worldwide in October 1989. The EPA was prompted into further research by the NRDC study and announced a ban from 1990 onwards because of the "inescapable and direct correlation" between its use and "life-threatening tumours". The British Ministry of Agriculture continued to give Alar the all-clear until Uniroyal withdrew it.[83]

The NRDC also raised concern over a group of widely used organophosphate fungicides – ethylene bisdithiocarbamates (EBDCs) – which it claimed are threatening some three million under-fives with damage to their nervous systems or behavioural impairments. Most concern has been expressed about ethylene thiourea (ETU), a break-down product of the EBDC fungicides. Children's exposure to EBDCs comes largely through tomatoes and tomato products, green beans, orange juice and cucumbers, among others.[84]

The EPA has classified ETU as a "probable human carcinogen" and threatened to ban or severely restrict the use of the EBDC fungicides. This prompted the major companies producing them voluntarily to rescind 39 of their 55 food uses in the United States in September 1989. Further studies are under way, and a final decision on uses and controls is expected from the EPA in spring 1991.

In Britain, a joint study by Parents for Safe Food and Friends of the Earth found residues of ETU in some foods bought from all major supermarket chains in the country. Samples included fresh and canned fruit and vegetables; products made from fruit and vegetables; bread; crisps and baby foods. Concentrations of ETU in certain foods exceeded the WHO's recommended maximum levels.[85]

The government's Advisory Committee on Pesticides completed its review of the suspect fungicides in January 1990 and concluded there was no risk to the consumer, although it conceded that ETU levels on lettuce and spinach were too high.[86] No restrictions on food uses for EBDC pesticides have been introduced in Britain.

Children are more vulnerable than adults to pesticides in food for several reasons. First, children eat proportionally more fruit and vegetables – and therefore more pesticide residues – than do adults. Relative to their body weight, children eat more of most foods than adults. Fresh fruit and vegetables make up about a third of the average US child's diet; the diet of a typical under-five is dominated by fruits, the foods most likely to be contaminated by pesticides. Relative to body weight, the average pre-school child consumes six times the amount of fruit an average adult woman eats, and 18 times as much apple juice.

Children's increased exposure to pesticides comes at a time when they are most susceptible to their toxic effects. Like lead, pesticides can affect the brain and central nervous system, which are still developing in the young child. Young children are more likely to retain a larger proportion of any dose of toxins because of increased absorption and decreased elimination. Young children absorb chemicals more easily than adults because the gastro-intestinal tract is more permeable and because the rate at which compounds are moved between cells is faster.

Young kidneys are also less capable of getting rid of poisonous substances. Enzymes which in adults act as detoxifiers are not fully functional at birth, which makes the young much less capable of neutralizing many chemicals than adults. Because of physiological immaturities, a child's organs are also more likely to be damaged by the poisonous effects of toxic compounds.

It is thought that children are probably generally more vulnerable than adults to cancer-causing agents. There are two reasons for this. First, cells divide more rapidly in infancy and early childhood, which increases the probability that cell mutations can be passed to subsequent generations of cells and can start the development of a cancer. Second, earlier exposure simply means there is more time for a cancer to develop.

The NRDC study showed that more than half the lifetime cancer risk that an individual faces from pesticide residues in fruits may be from exposures during the first six years of life.[87]

THE KILLING FIELDS

There are other ways for children to be exposed to pesticides apart from pesticide residues in food. Some, like contamination of groundwater and breast milk by residues, threaten children all over the world.

But many children in the Third World are particularly at risk if they or their parents work in the fields. Children in poor rural communities breathe in pesticides, get them on their skin, transport them from hand to mouth, and even get sprayed with them.

In Mexico over 40 per cent of the population is 15 or younger, and about 40 per cent of the population lives in small towns of less than 2,500 people devoted mostly to agriculture. Large numbers of children work in the fields, often supplementing the family's income, especially when the father has had to look for work elsewhere. Forty per cent of reported poisonings affect workers between 15 and 24 years old.

Women workers take their small children out with them into the fields and sit them between rows of crops where they spend the day breathing in pesticide residues, and in some cases being sprayed. Worker families living near the fields use empty pesticide containers for storing food and water and as drinking vessels.[88]

Vast quantities of pesticides – more than 20 per cent of worldwide sales in 1989 – are used in the Third World. The Third World's market for pesticides is growing faster than that of the industrial world. The industry maintains that it is misuse rather than recommended use of pesticides which presents the greatest danger to health. But conditions in much of the Third World are such that abuses are inevitable. Many pesticides which are banned in the West, with its complicated, sophisticated monitoring capability, are still on sale and in use in poor rural communities all over the world. Many "recommendations" are not written in the local language, nor could they be read by illiterate farmers if they were.

Governments North and South have much to answer for in their failure to regulate to defend the defenceless. In 1976, the giant Swiss drug company Ciba-Geigy purposely sprayed the pesticide Galecron on a group of Egyptian children to test its effects. The company later claimed the experiment was a unique mistake, but pointed out that it had the approval of the local Egyptian authorities. The children, who were paid for the job, had to stand in a field almost naked, while a plane flew overhead ıd released the pesticide. Urine samples were later tested to see ow much of the pesticide had been retained.

The scandal was discovered by a Swiss pressure group which ained access to the company's records with the names and ages f the children. Galecron was later linked with cancer and emporarily disappeared from the international market. Two ears later, it reappeared and was used legally for spraying cotton n several countries, including Mexico. In 1982, a Swiss television locumentary (which revealed the Egyptian scandal) alleged that Galecron was being handled by Mexican children.[89]

The problem of exposure in the fields is largely a Third World problem, but in all countries the children of farmers help their parents at harvest time. Despite regulations to protect farm-workers in California, pesticide poisonings have risen steeply since the 1970s. Reported cases of poisoning doubled between 1973 and 1985, and fieldhands suffer the highest rate of work-related illness in the state.[90]

WHERE ARE THE TESTS?

The real extent of the danger, particularly to children's health, from pesticides is not known, since the majority of pesticides used today have not been properly tested for health hazards.

In the United States, 390 of the 400 pesticides approved for food use are older pesticides which came on to the market before modern testing requirements were in place, according to estimates by the US government's General Accounting Office.[91]

In Britain, the first routine programme to review the safety of approved pesticides began in March 1989. By March 1990, only seven pesticides had been tested. Under the programme, pesticides approved before 1965 are reviewed first, but staff shortages are causing chronic delays; it is alleged that with 300 pesticides still to test, the process could take until 2033.[92] Before 1989, the only pesticides to have been reviewed in the light of new standards were those which had raised safety concerns. Of the 35 product reviews between 1964 and 1989, nine resulted in cancellations of approval and seven in restrictions on previous use; of the remainder many are still incomplete.[93]

Pesticides called in for special review because of concer about their safety, in Britain and the United States, are not take off the market. In the United States, special review normall takes between two and six years, but can take up to ten year Some 22 million American children can pass through their pre school years during the six years of a special review. Of th pesticides reviewed for the NRDC study, those which ar currently under special review pose up to 98 per cent of th cancer risk to the average pre-school child.[94]

DIOXINS: FROM FIRE TO FAT

Dioxins are by-products of burning almost anything. Chemica and waste incineration, power stations and leaded petrol are particular culprits.

Dioxins are a group of 75 chlorinated compounds, but the word "dioxin" refers in particular to the most poisonous of these, 2,3,7,8-Tetrachlorodibenzodioxin (TCDD), which is extremely toxic even in the minutest amounts.

Dioxins can pass into the body through the skin, or in food, or can be inhaled. They can cause cancer in animals and reduce the ability of the immune system to ward off infections, particularly in infant animals. They have also been linked with birth defects, stillbirths and sterility.

A British government report of 1989 was unable to set a definitive acceptable intake of dioxin for humans, although a guideline level was set at one picogram (a trillionth of a gram) per kilogram of body weight per day. The average adult's dose in Britain is just above the guideline level, at 1.3 picograms. But breast-fed babies are receiving average doses of 100 picograms per kilo of body weight through their mothers' milk – 100 times higher than the guidelines permit.[95] Other industrial countries have also found high levels of dioxin in human breast milk.

All fat-soluble and water-soluble chemical compounds pass into breast milk in significant quantities, because of the milk's high fat content. Human mothers are at the top of the food chain and accumulate these substances in high concentrations from the animal fats they eat. Body fat moves around in the body more during breastfeeding, and stored chemicals pass into the breast milk. As fat-soluble chemicals are very persistent in the body, and excretion in breast milk may be the only form of elimination, newborn babies are likely to absorb far more of these pollutants than their mothers ever did or than they ever did in the womb.

Infants have a hard time removing these substances from their bodies because their kidneys and liver are both poorly developed. Low-birthweight infants are particularly at risk because of this and also because the blood-brain barrier may be more easily breached.

These breast-milk pollutants, though extremely worrying, do not destroy the case for breastfeeding. Milk-substitute formula made from cow's milk also contains pollutants. For infants up to six months, human breast milk offers a perfectly balanced diet. It also transfers to babies the defences in the mother's body against the local bacteria and viruses. And breastfeeding seems to be important for the psychological well-being of both mother and baby. The advice of almost all doctors and women's health and environmental groups is that breast is still best. Mothers are increasingly involved in campaigning to get the offending chemicals out of the atmosphere, water and foods.

CHILDREN AND RADIATION

When President John Kennedy was trying to persuade the US Senate to ratify the Limited Nuclear Test Ban Treaty in the early 1960s, he said that he knew the numbers of children given leukaemia or born malformed because of the fallout from atmospheric testing might be small. But he argued that "the loss of even one human life, or malformation of one baby – who may be born long after we are gone – should be of concern to us all. Our children and grandchildren are not merely statistics towards which we can be indifferent".[96]

When Dr Adam Lawson, Chief Medical Officer at British Nuclear Fuels, was presented with evidence that children around the nuclear reprocessing plant at Sellafield were getting leukaemia because of their fathers' exposure to radiation in the plant, he pointed out: "It must be remembered that another 2,000 people working at the plant during the same period . . . have not got children suffering from leukaemia."[97]

British epidemiologist Martin Gardner looked at 97 cases of cancer in children born in the area between 1950 and 1985. The risk was found to be between six and eight times higher for children of fathers who had been exposed to a cumulative radiation dose of 100 millisieverts (mSv) or more by the time their children were conceived. There were also significant risks for fathers exposed to a much lower dose of radiation – 10 mSv or more – in the six months before the children's conception. Yet even this study could not make a case-by-case link between children's cancers and Sellafield, or say exactly how cancers may be caused in unborn children without their fathers being affected, or even how many cancers may be caused.[98]

One possible explanation of the Gardner study's findings is that radiation causes a mutation of some of the cells inside the testes which are copied to produce sperm; such mutations would produce large numbers of damaged sperm.[99]

Many of the villagers living in the area around the plant remained apparently unmoved, arguing that at least they knew what the scientifically calculated risks were, unlike those living near other industries. The village postmistress in Ravenglass went further, "A lot of healthy babies are born here. I think it must be good for you! There's a man who's 103 this year; Mrs Fox is 101 . . . and Mrs Preston was 95 last month".[100] There is little employment in the area, beside the reprocessing plant.

At high dose, ionising radiation will destroy any living cells through which it passes. At lower dose it can damage the genetic material within cells, leading to cell mutations and, ultimately, cancer or genetic defects.

The main effect of low-level radiation is on cell division, the process which allows living beings to reproduce themselves and grow. Since unborn children, infants and children grow faster than adults, each is correspondingly more at risk. When someone is exposed to high doses of radiation, injuries become obvious within hours. Lower doses can set off only partially understood chains of events which can lead to cancer 20 or 30 years later. If reproductive cells are damaged, the damage may be passed on to children, grandchildren or remoter descendants.

Safety standards have been set to protect the health of workers, the public and their children; these are meant to prohibit exposure above levels which could cause damage either to them or their reproductive cells. The current internationally recognized safety levels of 50 mSv for workers and 5 mSv for the public, in both cases over any one year, were set more than 20 years ago by the International Commission on Radiological Protection (ICRP).

In 1987, Britain's National Radiological Protection Board (NRPB) recommended a lower maximum level of 15 mSv for exposed workers; and that for members of the public, no more than 0.5 mSv per year of the permissible 5 mSv per year should come from a single nuclear site.[101] The board was anticipating pressure to tighten safety margins after new evidence from studies of cancers in Hiroshima survivors who had been very young at the time of the bombing.

The ICRP's standards have always been fiercely criticized. In 1988, Friends of the Earth "supported by 800 scientists from around the world" called for limits to be reduced to 10 mSv per year for radiation workers initially (with an eventual target of 5 mSv) and 0.2 mSv for the public.[102] That was before the Gardner report. The secretary of the ICRP has admitted that the revised standards it was due to publish in the summer of 1990 will have to be looked at again in the light of the new report.[103]

The most controversial finding in the Gardner report is the suggestion that long-term exposure to what had been considered very low doses – about 10 mSv – could be responsible for genetic changes that cause leukaemia. The implication is that safe levels can no longer be considered safe.

The Gardner report is not only important to workers in the nuclear industry and workers in other industries – iron and steel for instance – who may be exposed to naturally occurring radioactive substances, but also to the general public. Occupational safety standards have always provided the basis for public safety levels, as most of what is known about the effects of radiation on human beings comes from studies of men and women who have worked with it.

When radiation was used by only a small number of scientists and doctors, officials thought there was no need to set "safe" levels of exposure for the public. In fact, after the Second World War when it was first suggested that limits for public exposure to radiation needed to be established, the US Atomic Energy Commissioner declared that such a move would be "psychologically dangerous" because people would think any level of exposure was harmful.[104]

Today, the United Nations Scientific Committee on the Effects of Atomic Radiation (UNSCEAR) thinks precisely this. It bases its work on two rather frightening assumptions. First, there is no threshold below which there is no risk of cancer, and any dose, however small, increases the chance that the person who receives it will get cancer; every additional dose further increases that risk. Second, the risk increases in direct proportion to the dose received; if you double the dose, you double the risk.

Levels of safety aspired to in the nuclear industry are no more absolute than in any other industry. Attempts to limit radiation doses to workers and the public have always been bitterly contested, and they have always been set against the enforcement costs to industry and governments.

The purpose of standards is to make sure that the risks of exposure will be balanced by benefits: "All exposures shall be kept as low as reasonably achievable, economic and social factors being taken into account"; this is the ALARA, or "as low as reasonably achievable" concept.[105] But risks and benefits are rarely spread evenly throughout the population. The difference between safety levels for workers and the public is based on several premises. One of the most important is that, in theory, workers have the choice whether or not to work in a particular industry and are paid for the risk they take. But their children – all children – lack that choice.

Risk estimates differ wildly and are often more about economic and political pressures than about science or mathematics. As Michael Thorn at the ICRP put it in 1985: "Take the Third World, for example. It's not necessarily appropriate to spend the same amount of money on reducing radiation risks in an environment where there may be many other competing risks that are clamouring for a small amount of government resources".[106]

The Indian newspaper *India Week* has reported alleged cases of radiation poisoning near two of the country's largest nuclear plants. Villagers living near the plants are suffering from higher than normal rates of respiratory fever, miscarriages and cancers. The newspaper claims deformity and disease has increased in local children, including many cases of leukaemia and bone disease.

The Uranium Corporation of India says the plant emits less radioactive discharge than is permitted under UN regulations, but has refused to reveal to the newspaper the radiation levels to which workers are exposed. The Indian government does not allow the United Nation's International Atomic Energy Commission to inspect its nuclear plants.[107]

The Gardner report will encourage more scientific investigations in and around other nuclear plants in Britain and around the world. Within days of its publication, Britain's NRPB announced plans to amalgamate its own data on more than 100,000 radiation workers with data from the national childhood cancer register in Oxford.[108] The British Nuclear Tests Veterans' Association has called for an independent inquiry into levels of childhood cancer among children of servicemen who witnessed nuclear tests.

Many non-nuclear industrial plants – smelting and chemical works for instance – also discharge ionizing radiation into the environment, sometimes in sewers, but mainly into the air. The Capper Pass smelting plant near Hull, owned by Rio Tinto Zinc, is the largest metal-smelting plant in the world. It is licensed to discharge naturally occurring radioactive substances which arise from tin and lead production. Prompted by a high number of child leukaemias, brain tumours and other cancers in the villages around the plant, the local health authority commissioned an independent report from the Scottish Universities Research and Reactor Centre into possible links between the cancers and the plant.

The report, published in March 1990, could not find "a proven cause–effect link between Capper Pass radioactivity and local cancer incidence", but it was unable to "rule out the possibility of radioactivity from Capper Pass being a significant health-determining factor".[110]

In Britain about 200,000 cancers are diagnosed every year. Radiation is known to be a "causal factor" in childhood leukaemia, but many cases of childhood leukaemia have no apparent links with ionizing radiation. And there are too few such cases to provide the hard cause-and-effect links established – after many years and tremendous controversy – between smoking and lung cancer.

The doses of low-level radiation received by people living around nuclear plants cannot be measured in the same way that workers' exposure can be monitored. Effects of low-level radiation can take decades to show up, and genetic damage may not materialize for several generations. There is very little information available on the effects of radiation under 100 mSv. Until the Gardner study, many would have thought exposure even at that level would be too low to cause a measurable increase in an adult's risk of developing cancer. General risk factors have to be based on what is known of risk from high exposure, which is relatively easy to measure.[111]

The Gardner findings may increase the attention given to possible links between medical radiation and effects on children's health, both directly and through their parents' exposure. Most ionizing radiation – some 87 per cent – occurs naturally in the environment. Ninety per cent of all non-background ionizing radiation comes from medical exposure.

The risk of medical X-rays must be balanced against their benefits. Yet some experts are concerned that X-rays are administered too often and that doses for individual X-rays are too high and insufficiently controlled. Thirty-six million X-rays are conducted annually in Britain, the majority of them on teeth, chest and limbs. The dose equivalent for a standard adult chest X-ray is about 0.05 mSv.[112]

Some studies have shown that children are more likely to die of cancer if their mothers have been X-rayed during pregnancy, but these remain controversial. A 1986 study looked at 309 cases of childhood leukaemia in Shanghai. The disease was found to be more common in the children of men who had been X-rayed before the child's conception than in those who had not. The risk increased with the number of X-rays.[113]

Recently concern has been raised also about the possible health risks for children from some non-ionizing radiation. Research in the United States has shown evidence of disturbances in some children apparently caused by non-ionizing radiation emitted by major electricity transmission lines. Some schools have been moved away from the lines. At this stage evidence is inconclusive, but further research is under way.[114]

4
KEEPING CHILDREN ALIVE – FROM CALORIES TO CITIES

> Mankind is too apt to value things according to their present, not to their future usefulness . . . Upon no other principle is it possible to account for the general indifference to the death of infants.
>
> *William Buchan, 1769*

Pavitra, whose name means "pure", is 11 years old, but she left childhood behind some years ago. She is skinny, with a fine-boned face surrounding large black eyes.

She lives with her father and mother and two younger brothers in a tiny grass-roofed hut with a floor space of little over two square metres, in a village of such huts near granite quarries, about 50 kilometres south of Delhi in India. The hut is so small that when rains destroyed the roof, Pavitra was able to hold up one side of the new roof and help her father put it on. The three children sleep in the only bed; their mother, Lakshmi, and father, Sukhdev, sleep on the mud floor.

Every morning Pavitra rises at dawn, picks up two buckets and walks to the single village handpump to fetch the family's water. The water is not fit to drink because the shallow aquifer it comes from is highly polluted. Many pumps south of Delhi bear signs to this effect, but as there are no other sources of water, the signs are ignored. So many pumps have been installed that the underground water levels in the area are falling rapidly, and many pumps run dry. Pavitra hauls the water back, stepping carefully because there are no indoor toilets in the village, and everyone comes out in the morning to defecate on the ground.

Back home, she picks up a scythe and some twine and again walks off through the village. She stops briefly for a visit with her childhood, joining some girls in a game of "stones". It is played like marbles, but with small rocks instead of marbles. The game is punctuated by the exclamations of the girls, but the background noise is of sniffles and coughs from colds and sore throats.

She walks on about one and a half kilometres to a pit full of thorny branches, breaking them bare-handed, tying them into a bundle, and carrying them home on her head. Her mother scolds her gently for taking so long. Pavitra breaks the twigs into smaller bits to feed the fire carefully, so none of the fuel will be wasted. She cooks the flat bread chapattis on a clay stove out in the courtyard. The dried cow dung she adds to the fire makes smoke that is dangerous to her lungs, and her nose and eyes stream as she cooks. She hates the flames because a hut the family used to live in caught fire, and her legs were badly burned. She spent a month in hospital, the family bringing her food daily.

Pavitra has to do so much of the housework because her mother has been lucky enough to get a nine-to-five job in a detergent-packing factory, earning 18 rupees per day (about 60 pence). But the factory fumes make Lakshmi sick and feverish. She often has to take a day off to recover from a day worked. There is no sick pay.

Having fed her brothers, Pavitra picks up a hammer and walks with Sukhdev to the quarry. Her father passes her boulders of granite which she breaks into smaller pieces. It takes about four days to produce a truckload of small stones, and for each truckload the family is paid 150 rupees (£5) by the contractor and must then pay 10–20 rupees (30–60 pence) to the man from whom the small granite quarry is leased. Work slows during the hot season, when the stony ground radiates heat, and virtually stops during the monsoons. So there is often little money for food. This is not Sukhdev's chosen way of life, or even his native village. The family lived in the dry area of Rajasthan to the south-west, but drought made farming impossible and the family fled. Sukhdev could not find a job in Delhi, so settled in the quarry area.

The trucks carry the granite to one of 100 huge stone-crushing machines which break it down to gravel. These spew tons of dust into the air. The Indian Supreme Court has ordered the industry to take steps to limit the dust, but the order has never been enforced. Tuberculosis, chronic bronchitis and other respiratory diseases are rife in the area. The food Sukhdev's family eats always tastes of granite and earth.

It is a grim picture, but Pavitra has so far managed not to become a grim girl. She is very fortunate in a way. The quarries are full of hammer-wielding children as young as seven or eight who have been "bonded" into stone-breaking to pay their parents' debts. They have essentially been sold off and work for nothing. Pavitra lives with her family; she laughs, plays occasionally, and she has hopes. Asked about her future, she says she wants to go to college. Her brothers, she believes, will become car mechanics. She occasionally attends the "morning school" run for an hour a day by the local teacher.

It seems obscene to say that Pavitra is suffering from "environmental problems". She suffers from all the ills inherent in India's social, political and economic organization, and from India's place in the modern world. But these have been translated, for her and her family, into virtually all the world's major environmental ills: they are environmental refugees from the dried out Rajasthan area, now living amidst air and water pollution in a new, rocky, unproductive desert created by the stone quarries.

AN ENVIRONMENT OF HUNGER

Nutrition and the environment are linked in any number of ways. Spreading deserts and soil loss mean smaller harvests. Illnesses caused by bad air and bad water mean children's bodies use the calories they get less efficiently. Worms and other parasites, a natural part of the environment in many areas, may compete with children for the food they eat.

Nutrition is about much more than just amounts of calories and vitamins. Some nations do not grow very much food, in terms of their citizens' needs. But these are not necessarily the same nations which produce "wasted" children – children underweight in relation to their height. Rich countries, like most

Western European nations, produce very little of what their people eat but have very few malnourished babies.

India produces more than enough food to feed its huge population, yet over 30 per cent of all Indian babies in their second year of life suffer from wasting. The same is true of Egypt, a wealthier nation than India which produces ample food.[1]

Moreover, supplying the minimum calorie intake does not solve the whole problem. Experts are finding that even where the poor are eating as well as the rich, in terms of daily requirements, there are still more malnourished babies in the poor regions. Dr Alberto Pradilla, a Colombian who runs the World Health Organization's Nutrition Unit, argues this case forcibly. He notes that the distribution of energy intake, as a proportion of recommended levels, is about the same in parts of the United States as it is in Bangladesh and the urban slums of Cartagena, Colombia. Yet the latter two places have many more wasted babies. Why?

The answers lie in how nutrition is used inside the body. If a body is expending more energy in terms of calories than it is getting, then over time it wastes away. If it is getting more energy than it spends, then it stores fat. "It is just a balance between what you get and what you spend, like a bank balance", explains Pradilla. Obesity is just as much a form of malnutrition as wasting.

In the South, there are myriad reasons why ample deposits of calories and vitamins are not spent well in growth. Diarrhoea means that food is not absorbed from the gut. Worms may be eating food meant for the human body. Any disease burns up calories, keeps food from being absorbed properly and often causes diarrhoea or vomiting, which wastes calories.

"The best work on this was done in several villages in Guatemala", said Dr Pradilla. "The reasons for growth and lack of it were examined. Over 30 per cent could be attributed to diet; over 30 per cent to diseases; and over 30 per cent to unknown causes."

There are also apparent psychological and emotional effects on growth. Children taken into care, or put into orphanages, may be actually getting, and eating, an improved diet compared to what they were used to. But according to Pradilla, their growth often slows down.

TYPES OF HUNGER

Surprisingly little is known about malnutrition, considering that it afflicts more than 150 million children in the South (excluding China).[2] A pencil-thin Ethiopian child is obviously malnourished; so is an obese British child raised on a diet of sausages and chips; but the "hows?" and the "whys?" are much less well understood.

Part of the reason may be that no one ever dies of hunger, at least not officially. Malnutrition tends to make way for the disease that officially causes the death. The other part of the reason is that there is little funding available for studying nutrition among the poor. Much is spent on perfecting diets to help Europeans lose weight, but a lot less on the science of keeping weight on African children.

For example, there used to be wide agreement that there were two kinds of deprivation malnutrition. Generally shrunken children suffered from marasmus, a shortage of all types of food. The skinny children with swollen ankles, pot-bellies and bleached-looking hair suffered from kwashiorkor, a shortage of protein in particular. Yet recently, the causes of kwashiorkor have been called into question.[3] Mike Golden, of the University of the West Indies, has suggested that it may be caused by the body's reactions to infection. Ralph Hendrickse of the Liverpool School of Tropical Medicine believes it may be caused by fungal toxins in grain. The argument is not merely academic, in that if Golden is right the widely practised remedy of giving oil to children suffering from kwashiorkor may actually be dangerous. It is astounding that science is still so ignorant of the causes of such a widespread scourge at the end of the twentieth century.

Going by major symptoms, there are two basic types of malnourishment. Wasting is low weight for height; the children are forming insufficient muscle and not laying down enough fat. It can appear, and it can disappear, relatively quickly. Stunting – low height for age – is more chronic. Since it has to do with the growth of the skeleton, it takes longer to show itself, and takes longer to correct. It is associated with poor overall economic conditions, and represents the cumulative consequences of poor diet and a series of diseases over time. Wasting usually shows itself in the child's second year, after breastfeeding has stopped and the child is weaned on a poor diet. Stunting usually increases up to the second or third year and then levels off.[4] Children can suffer from both at the same time. But some areas tend to favour one form of malnutrition, and others the other. Scientists do not really understand why.

Stunting has an odd side effect, especially in areas where diets are improving. If a child suffers from stunting and then gets an ample diet, it may become obese because it is putting weight on a short skeleton. A national study in Chile of two- to five-year-olds found only 0.4 per cent suffering from wasting, but over ten per cent were stunted and 12 per cent obese.[5]

Another type of malnourished child is that which is born underweight. These are not premature babies, but babies which have gone full term and are born weighing less than the internationally agreed figure of 2.5 kilograms (5.5 pounds).

Out of 140 million babies born each year, some 22 million weigh less than 2.5 kilograms (kg): two million of these in the industrialized nations, more than 13 million in South Asia, and the remainder in the rest of the Third World.

Low birth weight is the single most important determinant of infant mortality, and usually reflects the health and nutritional well-being of the mother, so it is widely regarded as an indicator of the general social development of a population. Pregnant women's nutrition is determined not only by what they eat, but by the calories they "spend". If they are doing most of the farming work, and are walking long distances to fetch heavy loads of firewood and water, then they may be eating a diet adequate for a Northern sedentary mother-to-be, but still be malnourished.

Low weight-gain during pregnancy is related to low-birthweight infants. On average, Northern women gain 10–12 kg during pregnancy, while Southern women gain 2–7 kg.

A 1987 survey of the slums around Delhi found that almost 60 per cent of under-fives were malnourished, but so were over 80 per cent of all pregnant women and nursing mothers. Small, thin women given birth to small, thin babies, and risk their lives by giving birth. Every year at least 500,000 women die as a result of pregnancy (including abortions), 99 per cent of them in the South; and for every one that dies another 10 to 15 remain permanently handicapped.[6]

The link between infant mortality and low birth weight is proven. In New Delhi, a survey found that 238 babies died in their first year out of every 1,000 babies born weighing 1.5–2 kg; 59 died out of every 1,000 born in the 2–2.5 kg range; 21 in the 2.5–3 kg range and 18 for every 1,000 weighing over 3 kg.[7]

Low-birthweight babies are five times more likely to die of infectious diseases, three times more likely to die from diarrhoea and seven times more likely to die from respiratory infections.[8]

MOTHER'S MILK

Evolution has provided millions of years of product testing, which guarantee that mother's milk supplies all the nutrition an infant requires, in a convenient, sanitary container. This is true for the first four to six months of a child's life.

That, at least, is the accepted wisdom, based on controlled experiments which show that after that time, children do best if given some other foods to complement breast milk. Yet given the unsanitary conditions, bad water and often contaminated food in many Third World households, babies may continue to be better off if fed solely on breast milk, even beyond six months.

Mother's milk not only meets all of a baby's nutritional needs, but brings with it a package of disease preventatives. There are macrophages, which attack invading bacteria; there are immunoglobulins which coat the baby's gut with a material that stops disease-causing agents from getting through into the body. Breast

Opposite Mother and newborn baby in Burkina Faso, West Africa. Breastfeeding not only offers babies a perfectly balanced diet for six months, but protects them from the dangerous microbes in the local environment. *Mark Edwards/Still Pictures*

milk even acts against viruses, which few medicines do.[9] Since the baby is getting from the mother the best of the mother's immune defence system, the baby is also getting defences against locally occurring bacteria and other disease agents. Breast milk in Calcutta offers different protection from breast milk in Chicago.

Breastfeeding also makes mothers temporarily infertile. This is a neat trick on the part of Nature, because it ensures that the babe-in-arms gets a run of mother's milk and mother's attention before another baby comes along. When a mother is breastfeeding her baby often, when breast milk is all or almost all of the baby's diet, when the mother is not having monthly periods, then breastfeeding provides more than 98 per cent protection against pregnancy for the first six months after the birth of the baby.[10] It may not sound like a long time, but it is a crucial period for mothers and babies, and effective child spacing also contributes to controlling population growth.

Using infant feeding formula – breast-milk substitutes – raises many dangers for children, some obvious and some less so. The babies who do not get breast milk do not get the protection against disease it offers. Also, the formula itself may introduce into the baby's body bacteria against which it lacks protection. The milk powder may be mixed with dirty water, stirred with a dirty spoon, left open to bacteria in the air. Mothers may try to stretch the expensive powder by over-diluting it with water. If the baby does not finish it, the preparation may be saved until later; and few Third World homes have refrigerators. The bottle provides a pathway for the agents of disease in the surrounding environment to get into babies whose immune systems are not as well protected as if they had been breast-fed.

Even the fact that most pollutants find their way into human milk does not contradict the notion that breast is best. The levels of DDT and related pesticides in mothers' milk in many Third World countries is higher than international UN limits, but there has been no evidence that this has actually harmed children.

Bottle-fed infants are as much as 25 times more likely to die in childhood than infants who have been exclusively breast-fed during the first six months of life. During those months, supplementing breast milk with powdered milk can cause a tenfold increase in the risk of death.[11]

Since the bottle-feeding mother will be more likely to get pregnant, the baby may find a new sibling coming along in a year or so, a situation hazardous to a baby's life. Studies in Bangladesh have shown dramatic increases in the death rates of infants and toddlers when there is less than two years between children. One simple reason for this is that adequate spacing produces heavier, longer babies, more likely to survive.[12]

Given the almost miraculous qualities of breast milk and the dangers of substitutes, why do many women in the Third World insist on using commercial milk formula? Why is breastfeeding actually in decline in and around many Third World cities?

There are many reasons for this, some of them grim and nasty. But there is also a relatively positive reason. With more women being educated and achieving more social freedom, more women are going to work in offices and factories. But few get the sort of maternity leave offered in the North. And few businesses allow women to breast-feed on the job. Thus many such women make the reasonable choice to bottle-feed. It is a little unfair to have one set of "experts" telling women to take charge of their lives and use their education and get jobs, while another set of "experts" criticizes them for bottle-feeding.

But the decline in breastfeeding in Southern cities is also tied up with fashion and ignorance, as it was in Europe and North America when breastfeeding declined to about 30 per cent and below during the early 1970s. The ignorance is ignorance of the benefits of breast and of the dangers of the bottle; the "fashion" comes with the notion that substitutes, most of them made in the North or by Northern companies, are somehow more "modern", more civilized and therefore better for babies.

MARKETING MALNUTRITION

In one of the most cynical and unscrupulous business practices at work in the world today, many multinational food and drug companies are playing on mothers' anxiety and ignorance by encouraging mothers who can and should be breast-feeding to use the companies' powdered milk products. These are the same companies which sell foods and medicines in the North, and whose advertising campaigns strive to convince us that the firms have only their customers' best interests at heart.

In May 1981, governments belonging to the World Health Organization adapted the WHO/UNICEF "International Code of Marketing of Breast-milk Substitutes". Only one nation, the United States, voted against the code; three abstained; British delegates had fought hard for a tough code. The vote was a reaction against the carnage being caused worldwide by unnecessary bottle-feeding and the growing disgust at the "dirty tricks" the powder producers were using to sell their products.

The formula producers gave away free samples in maternity hospitals and wards. This is effective, because if mothers use such free samples for a few days after giving birth, their milk dries up; they cannot breast-feed and have no option but to continue to bottle-feed. The formula producers advertised directly to mothers; they hired women to give away samples and information, and dressed them as nurses. They were doing everything possible to catch their customers as early as possible and to associate their products in the mothers' minds with the best of modern medicine.

The code is tough in what it says, but weak in that it has no standing in law. Nations are expected to pass laws based on its articles and launch public education campaigns. Only a minority have done so.[13]

The code, among other things, forbids advertising of substitutes directly to the public; forbids the providing of samples to pregnant women or their families; forbids the use of health-care facilities – hospitals and clinics – for promoting formula or displaying products; and even requires that labels on substitutes display prominently a "statement of the superiority of breast feeding".[14] One key article says that donations or low-price sales of formula may be made to institutions or organizations, "but such supplies should only be used or distributed for infants who have to be fed on breast-milk substitutes".

Most Northern countries generally follow the advice of the code. Britain has its own weaker code, and in the United States companies have until recently had a sort of voluntary agreement among themselves not to advertise directly to the general public.

But in the Third World, where the media and citizen's groups are less powerful and less well informed, some of the manufacturers have continued to violate the code and exploit loopholes.

In 1988, the US-based pressure group Action for Corporate Accountability called for a new boycott against the Nestlé Corporation, better known for its chocolate and coffee but a leading producer of breast-milk substitutes. The group claimed that Nestlé was violating the code by giving supplies to hospitals and maternity wards. The new boycott was announced on the fourth anniversary of the end of a seven-year international boycott against the company over the same issues, a boycott which ended with an agreement on the part of Nestlé to abide by the code. Organizations in over 20 countries have announced support for the current action. A boycott was announced at the same time, for the same reasons, against the US-based firm American Home Products.

Nestlé argues that it is keeping to the letter of the code, and donating products for reasons of charity. Supporters of the boycott say that the powder is being "dumped" for promotional purposes and little of it goes to mothers who cannot breast-feed.

Speaking of Nestlé's activities, Dr Raj Anand, a professor of paediatrics in Bombay, India, said: "There is nothing 'charitable' about these so-called donations. The purpose of the industry in bringing the formula to hospitals is to induce sales. This is their one and only purpose."[15] A Filipino health worker, discharged from hospital with her new baby, complained: "I was charged for each cotton bud they had used to clean my baby's eyes and for the single teaspoonful of sugar they put in the water they gave her. But the Nestlé milk they gave her – against my wishes – was absolutely free."[16]

Alarmed at the dumping of promotional milk in hospitals, WHO's World Health Assembly passed a resolution in 1986 urging governments to ensure that hospitals and maternity wards buy the small amounts of breast-milk substitutes they need through normal channels and do not accept free or subsidized supplies. The formula companies argue that the resolution does not change the code, which allows for charitable donations.

The British Baby Milk Action Coalition (BMAC) regularly reports on what it sees as breaches of the code by companies. It accused the Milupa firm of West Germany and the My Boy firm of The Netherlands of visiting health-care facilities in Bangladesh and handing out gifts, toys and prescription pads with brand names, and free milk samples, which health workers passed on to mothers. The Coalition also charged that Farleys, a subsidiary of Britain's own Boots, has repackaged and added vitamins to three of its formulas and was marketing these in Britain directly to the public as a "follow-up milk".[17]

The manufacturers describe these follow-up milks as formula to be used after babies are weaned. Since they are thus not strictly a "baby-milk substitute", the companies say that they do not fall under the articles of the code. Yet both UNICEF and WHO have described these products bluntly as "not necessary".[18]

The UN does not criticize companies by name, but its latest report on the implementation of the code includes complaints of violations from many countries ranging from rich New Zealand to poor Ethiopia, and including Colombia, the Dominican Republic and Bangladesh. It would seem a low trick for companies to take advantage of Ethiopia's poverty and administrative muddle, but the Ethiopian government complained of "the proliferation of brands of infant formula, most of which failed to respect the labelling provisions laid down in Article 9 of the international Code" (the part of the code which, among other things, calls for a notice on the label that breastfeeding is superior).

WHO bases its annual reports on government data, and few governments bother to report violations of the code. But Colombia complained rather plaintively that despite its adoption of strict standards to control substitutes in 1980 "the promotion of these products is still carried out in almost all maternities [maternity wards], and free samples are requested and accepted in most hospitals and clinics".[19] Companies that Europeans and North Americans are doing business with every day continue to profit from the deaths of the youngest and the most innocent.

And things may get worse. Bristol-Myers Co. has 35 per cent of the $1.6 billion dollar US baby formula market. In 1989 it announced plans to advertise directly to the US public a new formula, to be marketed by Gerber Products Co. This led to calls for a boycott, not from pressure groups, but from doctors across the country. For doctors to speak out against Bristol-Myers, a company which provides many perks for doctors, including scholarships, reveals a deep outrage. A Nestlé subsidiary had begun direct advertising of two new formulae in 1988, but stopped advertising one product when the American Association of Paediatrics criticized the action. It continued to advertise the other, a "follow-up milk".[20]

These US wrangles do not directly affect the rest of the world. But officials at WHO in Geneva have been collecting clippings on the fracas, some expressing fears that what self-restraint US firms have been showing may be cracking, and this could eventually mean more aggressive marketing overseas.

UNICEF is outspoken on the causes of the problem: "Breast-feeding appears to be on the decline in many developing nations as commercial pressures, the use of milk powder and feeding bottles in hospitals, and the increased participation of women in the labour force all conspire to make bottle feeding seem an attractive option. The continuation of this trend would be disastrous."[21]

The UN tends to understate and dodge many issues; on this problem it is becoming ever more outspoken. Three UN agencies – WHO, UNICEF and UNESCO – recently published a booklet entitled *Facts for Life*. On the breastfeeding issue it contains the simple sentence in bold, blue ink: "Bottlefeeding can lead to serious illness and death".[22]

That must be a difficult statement for the executives of formula-manufacturing firms to live with.

CHEAP FIXES

Millions of families cannot grow or buy enough food to feed their children. However, most families could give their children an adequate diet if they knew which foods to provide. Education and a few cheap government interventions can keep children healthy and protect them from some nutritional scarcities in the local environment.

Most children who suffer wasting begin to do so early in their second year when first weaned on to solid foods. Many parents tend to fill their children's very tiny stomachs once or twice a day with a local staple. Depending on what the staple is, the children may be full, but not getting enough protein. Cassava and bananas are particularly high-bulk, low-protein foods. Children under the age of three need feeding five or six times a day: the normal porridge – usually based on the local grain staple – but enriched with mashed vegetables and a very small amount of fat or oil.

Weaning has always been a dangerous time for children. Typical weaning foods in Victorian Britain included mixes of arrowroot, oatmeal, sago, cornflour and baked flour. But before an infant was one year old, it got whatever was left over from the family table, which means that many Victorian babies were weaned on beer, cheese, onions and heavy breads, not a good diet for a small stomach and a delicate digestive system.[23]

One of the cheapest ways in which mothers can monitor whether their children are getting enough to eat is to weigh and measure them regularly, comparing the results against the norms on standard charts. If the child is far below the norm, mothers can seek advice or medical help, if it is available. This growth monitoring has become a cornerstone of much of the work with infants done by the United Nations and the development agencies.

While we tend to glamorize folk wisdom, some of it can be deadly. Many Third World parents stop feeding children who are ill, believing this is best for them. The belief is reinforced because ill children often have little appetite. But fevers and the body's defences against infections burn calories, so ill children, more than most, need frequent, small doses of easily digestible food.

The vitamin deficiency most permanently crippling to a child is that of Vitamin A, the lack of which blinds 200,000 children every year. A handful of green leafy vegetables a day can prevent loss of sight. Other foods, such as breast milk, carrots, papayas and mangoes are all rich in Vitamin A, which also plays a role in preventing diarrhoea. Unpublished research in India and Indonesia suggests that Vitamin A supplements can also cut infant mortality, by perhaps as much as 60 per cent, though no one understands quite how, and such results need further research.

Iodine deficiency is very much an "environmental" condition. At one time, virtually all the Earth's soils were rich in iodine, which got into foods and drinking water and thus into the evolving human species, where it became part of the thyroid hormone. But gradually rain-water washed iodine out of many mountain soils, and frequent flooding took iodine out of the soils around the estuaries of tropical rivers – and hence out of food crops. Lack of iodine swells the thyroid gland, producing a large growth at the front of the neck known as a goitre. But iodine deficiency also means that the thyroid does not do its job, resulting in retarded physical development and impaired mental functions. Spontaneous abortions and still births increase. Babies of mothers who have had insufficient iodine may be born as cretins; as children, they may suffer dwarfism and severe mental retardation. They may also be deaf mutes.

Eight hundred million people live in areas of the world, found on all continents, which are short of natural iodine. The industrialized nations have solved the problem by adding iodine to table salt, and a few have iodized bread and water as well. So there are only a few pockets of central and southern Europe left where people are prone to goitres. But iodizing salt has proved more difficult in developing nations, where people get their salt from so many different sources, not all of them commercial. Nations such as Indonesia, Peru, Zaire and even New Zealand have successfully coped with the problem by injecting people with oil containing iodine, which gives them protection for about four years. However, throughout the Third World, an estimated 190 million people have goitres and 3.25 million suffer cretinism.[24]

UNICEF has listed Vitamin A and iodine deficiencies among what it calls the "passive atrocities", because preventative measures are so well known and so cheap. It reckons, perhaps a little naively in the case of some large nations in which the population is scattered widely across the countryside, that iodizing all edible salt would cost less than five cents per person per year. David Haxton, former UNICEF director for South-central Asia, said upon his retirement in 1989, after wrestling with the problem throughout his career: "Permit me to suggest that it is a crime for one more child to be born a cretin. We have known the answer to prevention for over 75 years. . . . Must we end this century with hundreds of millions still at risk when we know the answer and can afford the price?"[25]

Malnutrition remains a problem among older children in the developing world, where 40 per cent of all children between the ages of 6 and 14 are underweight (no figures for China). This lack of proper growth stems from all the usual reasons associated with poverty and ill-health. But in cities "junk food" is an increasingly important factor as children get older. Food sold by street vendors rarely comes under the scrutiny of public health officials. It is often prepared in dirty surroundings, stored warm in the open air, and contains questionable additives and colouring agents. These risks, on top of the general lack of refrigeration and hygiene in many homes, means that diarrhoea, a killer of infants and toddlers, continues to be a problem among older children. Dr Steven Esrey of Johns Hopkins University estimates that as much as 70 per cent of all Third World diarrhoea, normally associated in young children with bad water, may be caused by food contamination.

DIARRHOEA: THE PREVENTABLE KILLER

Diarrhoea and associated diseases kill four million children under the age of five in the Third World each year. But diarrhoea itself is not so much a disease as a healthy response on the part of the human body, as it tries to flush out harmful substances in the gut.

This may be a reaction to viruses, as in gastro-enteritis; to bacteria, as in bacillary dysentery and bacterial food poisoning; or to parasites, as in amoebic dysentry or worm infestations; or even to drugs, poisons and foods to which the body is allergic.[26] Normally, the wastes in the small intestine are always watery; the large intestine is supposed to absorb most of that water. This action makes for a regular stool and supplies the body with fluids which it needs to survive. Thus the major cause of death in many diarrhoeal diseases is loss of water and loss of potassium and other salts.

About 2.5 million of the 4 million diarrhoeal deaths every year could be prevented if the children's bodies could only be kept topped up with sufficient water while whatever is causing the condition runs its course. (Preventing the other 1.5 million deaths would require other treatment, as well as the replenishing of fluids.)

The high-technology way to get water into the bodies of badly dehydrated children, and adults, in the North is to stick a needle into a vein in the arm and drip liquid in slowly. This is fine if the patient is in hospital and has the benefits of skilled nurses and the right equipment (bags of the solution, stands for the bags, tubing and a sterile needle). But it is obviously not helpful in an African hut or a Latin American shantytown. Even in the field hospitals of the Ethiopian famine camps, intravenous drips pose problems. One of the saddest sights in the world is that of a nurse probing the wasted arm of a dehydrated one year old to find a tiny vein in which to insert a needle for a drip. There is rarely enough skill or equipment to cover all those in need.

There is a simpler answer. If a teaspoon of table salt and eight teaspoons of sugar are mixed in a litre of water, a solution is created which passes more easily than plain water through the child's intestinal lining. The salt replaces lost salts, and the sugar helps to carry the water into the body. This is not a new treatment; it has been around for centuries in the form of mixtures of rice water, soups, fruit juices and water from cooked cereals.

But the sugar-salt mixture, now commonly known as "oral rehydration therapy" or ORT, allows parents to get the mixture just about optimum, so that the child gets the most benefit. The old remedies tended to be hit-or-miss affairs.

ORT shot to international medical prominence in the 1970s because it offered a simple, home-made answer to the problem of most diarrhoea, then killing more than five million children each year. Campaigns were mounted and instructions were handed out in the form of "pinch of salt and handful of sugar to a litre of boiled water". Mothers in Ghana were taught to measure the ingredients into a standard-sized, nationally available soft-drink bottle. Measuring spoons were issued in many countries.

Since then, UN specialists have become a bit nervous, not about the life-saving qualities of ORT but about mothers' abilities to make up an effective mixture at home. Many homes lack "sugar" in the form of a purified foodstore variety and may have a sweetener like honey or molasses instead. Or they may not have measuring spoons. If too much salt is added, then the solution can have the opposite effect and draw water out of the stomach lining. One book on sale in Britain has the mix reversed, describing it as a handful of salt and a pinch of sugar.

To avoid these problems, WHO and UNICEF have organized the manufacture of pre-mixed sachets of ORT salts and sugar in most countries for distribution to health centres and villages. These guarantee that the proportion is right, if the right amount of water is added. They contain sugar more easily digestible than sucrose, and potassium as well as the sodium in table salt (sodium chloride), so the right salts are replaced.

Each sachet is produced at a cost equivalent to ten US cents. UN officials place their hands on their hearts and swear that this move to manufactured sachets does not represent the greedy commercialization of a simple, home-made product, but is actually best for the children. In the refugee camps in Africa, health workers still tend to make their own huge batches of ORT solution, as their hands get too tired from ripping open the individual dose sachets, even if they have been delivered to the camp in the first place.

ORT is now saving the lives of one million children each year, according to UN estimates. But 2.5 million children who could be saved by ORT continue to die annually. WHO estimates that in 1988, based on evidence from 46 nations, only 14 per cent of doctors, four per cent of nurses, eight per cent of paramedics and nine per cent of community health workers had been trained to use ORT, despite unanimous expert acknowledgement that it is "potentially the most important medical breakthrough this century". The key question facing governments in 1990, according to UNICEF, is "whether or not the obvious thing will be done – will ORT be made as available and as well known as Coke or Pepsi or will we watch 25 million more children die of dehydration in the decade ahead?"[27]

THE DRUG PUSHERS

There are several hurdles in the way of doing the obvious thing. First, it is hard to spread the facts to health workers and the public in parts of the world where communications are poor. Second, professors of medicine do not like to spend years at school only to find themselves recommending simple sugar and salt.

Another barrier to saving lives is the very medicines being sold to prevent diarrhoea. A high proportion of these are made and distributed by the Northern multinational drug companies and their subsidiaries, but local drug companies also manufacture and sell their own brands in developing countries. These medicines have many drawbacks.

First, they do not work. Some have a modest effect on the duration of the diarrhoea but do not reduce the loss of fluid and salts which leads to dehydration. In addition, they divert attention away from the need to replace fluids. They can be deadly for children if parents use them in place of ORT. Some so-called anti-diarrhoeal medicines have serious side-effects.

But advertisements for the commercial drugs claim they stop diarrhoea. ORT is less well advertised, and it does not stop the obvious symptom of diarrhoea itself. Instead, it keeps children alive.

Debbie Christie of Granada Television in Britain took her cameras to Kenya to document the marketing of anti-diarrhoeal drugs there. She found ORT sachets costing six pence a packet saving lives in clinics, while at chemists' shops parents were spending the equivalent of £2, a huge proportion of their weekly income, for a bottle of an advertised medicine. More than half the mothers visiting the clinic with dehydrated babies had tried the more expensive drugs before coming in. Also, the mothers had sold food which should have been given to their sick children to be able to buy worthless medicine.[28]

The first drug discussed was ADM, manufactured by the British Wellcome Foundation and widely advertised in Kenya. Like many such medicines, it contains kaolin, a fine clay used in making porcelain, and pectin, a jelly-like carbohydrate used in setting jellies and jam.

According to WHO, such products do nothing to shorten the duration or reduce the severity of diarrhoea, or reduce the loss of salts and water. In fact, all they do is give a child's stool more shape and form. Soon after the programme was screened in Britain, Wellcome announced that it was withdrawing ADM from the market worldwide. "If you are one of those people who wrote to Wellcome, congratulations on some effective campaigning", said Christie.

Other products remain on the Kenyan market and are still widely advertised. Imodium, promoted by Janssen, a subsidiary of the multinational Johnson and Johnson, is loperamide, a chemical which paralyses the gut and stops the spasms of diarrhoea. WHO notes that it can have serious effects on young children's nervous systems; it adds that there is no rationale for the production and sale of liquid and syrup loperamide medicines for children. In Britain it is not allowed for children under four.

Antibiotics are also sold for the treatment of diarrhoea in much of the Third World, as in the product Kaomycin, manufactured by the US firm Upjohn, and containing kaolin, pectin and the antibiotic neomycin.

Antibiotics are not recommended for the routine treatment of diarrhoea, and can change the local "germ environment" by producing bacteria resistant to antibiotics. This means antibiotics become less effective in the treatment of other diseases against which they are normally effective. WHO says that the use of oral neomycin can actually prolong diarrhoea and make it worse. It "should not be used in the treatment of diarrhoea. The production and use of these 'antidiarrhoeals' cannot be justified."

The campaigning group Health Action International believes that combining antibiotics with diarrhoea medicines is "a foolish, dangerous and useless practice: most diarrhoea is caused by viruses which do not respond to antibiotics. Yet some 65 per cent of antidiarrhoeal preparations on the market in five regions of the world in 1985 contained an antibiotic."[29]

In welcoming the decision to withdraw ADM from the market, the British medical journal *The Lancet* commented that both Imodium and Kaomycin "are unacceptable for the treatment of childhood diarrhoea". Imodium "is an effective drug of convenience for adults, but the wrong message with it, promoting its use in small children, is dangerous and indefensible". Kaomycin "contains neomycin – an active drug but irrelevant to the treatment of diarrhoea and capable of serious toxicity".[30]

Christie and her team found that the dangerous, expensive and inappropriate diarrhoea drugs are muscling out ORT where it is needed most: "In southern India, 86 per cent of children receive anti-diarrhoeal drugs, only seven per cent ORT. . . . In Africa, 57 per cent are given drugs, only 12 per cent ORT."[31]

WHO is preparing a report on the drawbacks to these and other drugs, which it plans to publish in 1990.

Diarrhoea is rarely a threat to the lives of healthy, well-nourished children in the North. Most chemists in Britain recommend and sell inexpensive ORT packets for infant diarrhoea, though some still recommend the more expensive kaolin/pectin compounds.

WATER, PEOPLE AND "THE DECADE"

The importance of water in treating diarrhoea suggests how crucial water is to human life in general. But poor water is life-threatening, and much of the planet's population drinks poor water. ORT, effective as it is, should be seen as only a temporary solution until most of the world's children are provided with water safe to drink.

In the late 1970s, as scientists and doctors were collecting data to prove the need for a world-wide assault on unsafe water, the *New England Journal of Medicine* published a league table of all water-related illnesses and their effects.[32]

Diarrhoeal diseases were estimated to be killing between five and ten million people per year (adults and children); schistosomiasis (carried by snails living in slow-moving water) was killing 500,000–1,000,000 people yearly; river blindness (onchocerciasis), carried by blackflies associated with fast-moving streams, caused 30 million infections per year. Hookworms were causing 700–900 million infections per year, with an average of 100 working days lost per infection. Roundworms were causing 800 million to one billion infections, with 7–10 days lost per infection. Malaria, carried by mosquitoes which need water for breeding, was estimated to be killing 1.2 million people yearly. It is now estimated to be killing two million annually, most of them children. The list goes on with its millions, including polio, typhoid, and amoebic infestations.

Such figures are incomprehensible in terms of human suffering, but make an unanswerable case for any efforts to bring safe water into or near homes, and to limit the number of people collecting it from streams, lakes and puddles.

Dr Sandy Cairncross, of the London School of Hygiene and Tropical Medicine, points out that in many regions of the Third World, especially the drier Third World, women fetching water have to walk more than a kilometre carrying a full container weighing more than a suitcase for which an airline would charge excess baggage rates. They must usually walk uphill, as water is found on low ground. This work cuts into the amount of time and energy a woman has for preparing food, cleaning the house (thus making it healthier) and taking care of the children. A woman breastfeeding may suckle her baby 50 times in a 24-hour period, and cannot satisfy the child's hunger if she is rushing off to fetch water. There is less water available to wash children's hands and faces. City families without taps or nearby free pumps may spend one-fifth of the family income buying water from vendors, leaving less money for food, fuel, health care and education.[33] Providing clean water can improve the lives and survival chances of children in ways which go far beyond the harmful effects of bad water on children's bodies.

With such facts in mind, the UN system launched in late 1980 the "International Drinking Water Supply and Sanitation Decade". Its goal was nothing less than to bring safe water and adequate sanitation (the removal of human wastes) to *all* by 1990.

The Decade, officially ending in late 1990, "failed" in that it did not meet its "for all" goal. UNICEF estimated in 1988 that, since 1980, safe water had been made available to an additional 700 million people and sanitation to an additional 480 million. These impressive figures become less impressive when population growth over that period is factored in. In the Third World 60 per cent of rural families still lack safe water, and 85 per cent go without adequate sanitation. In the cities, 23 per cent lack safe water and 42 per cent acceptable sanitation.[34]

The reasons for the failure are many. The Decade was "co-ordinated" by several UN agencies, but effective co-ordination did not begin until towards the end of the decade. Key Northern governments, locked into free-market ideologies, showed little interest in the effort. The Third World nations suffered from rising debts and falling commodity prices throughout the Decade. And when the Decade started, the drills and pumps used were inefficient and tended to break down often; most of the technology was expensive. Now, as one UN official put it: "We learned enough during the Decade to mount an effective Water and Sanitation Decade – just as the Water and Sanitation Decade is ending".

Plastic pipes are now cheaper and more effective. Drills are lighter, cheaper and more mobile, and can drill a well through rock in a day where early-Decade drills required a year. New pumps break down far less often. Cheap, often waterless toilets and latrines have been developed to meet most of the world's social and environmental conditions.

But the most important lesson of the Decade was that water technology cannot simply be delivered to a village by outsiders of another race and culture – even by outsiders of the same race and culture – and the villagers then left to get on with it. As will be seen, this is true of all forms of technology, even the simplest techniques to improve health.

The villagers who will use the water and sanitation projects must help to plan them; they must be involved in choosing the place for the pumps and latrines and in constructing them, and they must be trained and encouraged to organize to maintain and clean the facilities and get the spare parts themselves. This sounds obvious, and becomes even more obvious when one imagines what the results would have been if Nigerians or Indians had decided where to place the pumps in English villages a century ago. But working with local people is more expensive and time-consuming, and it took innumerable failures before aid agencies and governments learned that it was the only path to success.

Millions of people are now waiting to see if governments can find the political will for another "Decade", to meet the new UN goal of safe water and adequate sanitation by the year 2000.

MAKING DESERTS FOR CHILDREN

There is another process at work in the world today which has a direct effect on children's ability to get enough food and water, to stay healthy and to earn a living when they grow up. It is often described as a "global threat", but its worst effects are local, destroying the livelihoods of the poorest families in the poorest countries.

The droughts and hunger of the early 1970s in Africa's Sahelian region – that band of nations running across the continent south of the Sahara – focused world attention on what was happening, to a greater or lesser extent, in all of the planet's dry regions.

Scientists collected data, and lumped the alarming trends under the unhappy word "desertification". The word is unfortunate because it has never been properly defined. The UN system tended to use it to cover various types of local over-use and misuse of the environment in dry areas, all of which left once productive land unproductive. These crimes against the land included over-cropping, over-grazing, deforestation and wrong-headed attempts at irrigation which spent a great deal of money to create wet, salty, sterile fields.

The UN Environment Programme (UNEP) has collected the usual set of ungraspable global figures: an area of land equal to the size of North and South America combined is suffering loss of production due to desertification; every year an estimated six million hectares of productive land go out of production, and another 21 million become so degraded that crop production becomes uneconomic; the problem was affecting 57 million people in 1977, 135 million in 1984, and the numbers continue to grow.[35]

Many scientiest remained sceptical. They did not like vague terms like "degraded"; they did not trust the figures; they did not know to what extent over-farming and over-grazing could cause *permanent* damage. Today the word "desertification" seems to be going out of fashion. The biggest problem with it is that causes cannot be separated. Which are local? which regional? which global? and which may be connected to some sort of long-term "natural" climate change which no one understands?

A recent report on future prospects for the Sahel could not determine whether the drying out of that region over the past several decades was part of a trend going back 8,000 years, or was some sort of rupture in the trend leading to accelerated dryness. Or is it being caused by the greenhouse effect and general "global warming"?

The writers of the report, associated with the group of donor nations called the "Club du Sahel", focused on the fact that much of the tropical rainforests of West Africa have been cut down this century. The Ivory Coast had 15 million hectares of such forests at the beginning of the century, but little more than three million in 1985. These wet forests hold a reservoir of moisture; as they disappear, there is less moisture available for the south winds to carry northwards over the Sahel. They concluded that this hypothesis – plausible but unverified – is the most likely cause for the recent accelerated drying of the region.[36]

If the scientific causes of desertification in the Sahel, and elsewhere across Africa, Asia and Latin America, are unclear, the political causes are much more obvious and much less frequently discussed. In the industrialized nations, farmers are the minority, but have much more political power than their numbers merit. Drought and soil degradation in the dry areas of the United States and Australia induce an outpouring of government relief from national, state and local governments. There are nationally subsidized crop insurance schemes, technical advice, new dams for irrigation, etc. Taxes from the majority go to support the minority.

Not only did US farmers not die in the Midwest and Southern droughts of 1988, but farm incomes actually improved because those hardest hit received generous disaster relief, and those crops which did grow drew high prices in the commodities markets. In the drylands of the Third World, farmers and their families are the majority, but have much less political power than their numbers should warrant. The poorest farmers live in the arid regions, on the political and environmental margin. Most of the cattle-herding nomads of the Sahel belong to tribes that are racially and linguistically different from the majority of the national populations; these tribespeople are rarely represented in the governments in the capital.

For poor farmers, global trends have little meaning. The livestock they lose to drought are lost from an already small, weak herd; the crops that die are crops from small, poor plots. Women and children must go further to find firewood and water. There is less wood to cook less food. There is less water for drinking and washing. Desertification can affect almost every aspect of a child's health and nutrition.

Though the causes of desertification and the causes of malnutrition are many and varied, a chart of the African nations with the highest percentages of malnourished children (compiled before the mid-1980s drought and famines) corresponds almost directly to a chart of the countries with the worst desertification. Burkina Faso, Ethiopia, Chad, Mali, Angola, Mauritania, Somalia and Niger (in that order) are in the top 15 nations with the most malnourished children. Some 40 per cent of the children in thinly populated Burkina Faso were underweight. In much more populous Ethiopia, more than 37 per cent of the children were malnourished, a total of 2.2 million children.[37]

Desertification can also affect governments' and agencies' attempts to supply people with safe water. Much of this work has concentrated on installing simple, effective, easily repaired handpumps. Yet even the best pumps, and even the most committed villagers, can only pump up water when the water exists.

The Indian government was deeply shocked by its 1984 survey of progress toward meeting the goal of safe water for all by 1990. It found that in the Uttar Pradesh northern hill district, out of 2,700 village drinking water supply schemes, 2,300 had failed as the water sources had dried up or sunk below the depth of the pumps.[38] Various reasons were offered, including increasing water run-off caused by hillside deforestation, and simply too many pumps used by too many people. Such findings may be a sign of things to come elsewhere.

Governments have been unable to find much reason to care

about desertification. UNEP held a global conference on the problem in 1977. The resulting "Action Plan" required them to hold another meeting in 1984 to examine progress. In preparing for this second meeting, UNEP sent out elaborate questionnaires to all governments. Few governments in the Third World had the data with which to fill them out; so UNEP hired consultants to help them. Still the questionnaires remained incomplete. The governments just did not know how much of their land was being turned to desert each year, or where, and how many of their people were affected, or how.

True, many of the worst affected countries lack experts to do this sort of survey. But few had chosen to use foreign aid to help them either study or tackle the problem. After the first Sahelian drought, aid money poured into the region; but only four per cent was spent on helping farmers to grow food on unirrigated land, and only 1.5 per cent was spent on tree-planting and soil and water conservation, or on what can be considered anti-desertification work.[39]

UNEP estimates that, globally, desertification causes agricultural production losses worth $26 billion per year; spending $4.5 billion per year would bring the problem under control. There are clearly big profits to be made in taking action. The drawback is that much of this profit would go to the poor and the powerless. So the money is not spent.

The 1988 Club du Sahel report notes that if the loss of rainforest is causing the drying of the Sahel, then "the next quarter-century will be markedly drier on average than the last quarter-century, with even more marked dry periods – perhaps so marked as to constitute major catastrophes".[40] This vision does not take into account global warming and its possible effects on the region. If catastrophes do emerge, they will be caused as much by political choice as by climate change.

THE VANISHING SEA

The Soviet Union is losing a sea. It is one of the most sensational examples of human-caused desertification on earth, and also one of the best examples of how, in issues relating to both children and the environment, everything is related to everything else.

In 1960 the freshwater Aral Sea in Soviet Central Asia covered 67,300 square kilometres; today it covers 40,400; in 30 years it could be gone. In an attempt to grow cotton in the surrounding dry area, the Soviet government established irrigation projects which took huge amounts of water out of the Amu Darya and Syr Darya rivers, the only rivers feeding the sea. The Amu Darya now carries no water to the Aral. The irrigation works leached salts from the soil and deposited them in the sea, which once held 24 native species of fish.

It now holds none, and the big boats which used to fish the sea lie aground on the salty desert far from the present shores. The fishing port of Muynak is now over 30 kilometres from the water, and its fish processing plant processes fish flown in from as far away as the Atlantic.[41] Wildlife species in the region have decreased from 178 to 38.

As word of this major change to the landscape leaked out to a Western audience, it was seen as a major "environmental" blunder, or series of blunders. But for the local people it is a major human tragedy that is killing a people and its children. The Soviet newspaper *Literaturnaya Gazeta* reported in mid-1989: "On the banks of the Aral Sea looms a catastrophe to be compared with Chernobyl and the Armenian earthquake. Infant mortality is one of the highest on the planet. Epidemics rage there."

There is the loss of earnings from the fish and the lost livelihood of the fishermen. There is also the loss of farmers' earnings as the government attempts to put water back into the sea by reducing irrigated cotton cultivation in the region.

The storms in the area pick up the salty dust in the new 26,900 square kilometre desert, land that was once under water. This dust covers fields far away, and it gets into the eyes and throats of the people. Throat cancers are increasing rapidly in the older people, eye and respiratory diseases in the younger. More and more anaemic infants are being brought to the local health centres.

The summers are much hotter than before. The watertable is falling, so wells are running dry, and all of the childhood diseases associated with lack of water are increasing. Infant mortality rates are now the highest in the Soviet Union, over 100 per cent higher than the national average. In the former port of Aralsk, one in ten babies die in their first year; there are almost 30 times more typhoid cases than previously, and hepatitis cases have risen sevenfold, according to a Soviet expedition to the area in 1989.[42]

Dr Kairbai Todimoratov, a physician in the former fishing port of Muynak, says simply that 83 per cent of all the adults are "sick", suffering everything from gastritis and hepatitis to typhoid.[43]

The cotton farmers in the area were using 52 times more pesticides and herbicides on their crops than are used in the United States, and many of these were poisons banned in the West. These too are blown about in the winds and inhaled with the dust. They have contaminated the water supply and mothers' breast milk. Many women have been warned that their own milk is poisonous, but they have little choice but to continue breast-feeding.

This is not only desertification on a spectacular scale, but a sign of things to come in other parts of the world, according to Rosbergen Reimov, director of the local academy of sciences: "This is a kind of a preview of global warming. The whole world should take note."[44]

FORESTS AND THE FUTURE

The loss of forests is one form of desertification, and another example of short-sighted land use which hurts the livelihoods of children today and threatens the welfare of future generations.

Forests are being cleared by farmers and ranchers, many of them subsidized by governments. They are being felled by timber companies, most of them encouraged by governments, which get part of the profits. Most of the governments in question are treating their forests as drunken gamblers treat their bankrolls – as an inexhaustible source of wealth. A recent survey of most of the tropical moist forests which are being logged found that far less than one hectare out of every 800 hectares of productive forests was being managed for sustainable production.[45] But open woodlands are also being lost at rates which deny millions firewood for cooking and heating.

Wood is the third most valuable primary commodity in world trade, after oil and natural gas. So the waste of forest is a waste of national fortunes.

Any figures given for the rate of forest loss are almost certainly wrong, because no one really knows what is happening in the forests, and because definitions of "deforestation" vary. The UN reckons that between 1964 and 1984 about 13.1 million square kilometres of the world's closed forest and more open woodlands were cleared – an area considerably larger than all of China.[46]

Estimates of the loss rates of tropical moist forest – the rainforests and "jungles" of Latin America, Asia and Africa – vary considerably. The usual estimate of 61,000 square kilometres lost annually, an area somewhat smaller than Ireland, is based on a very strict definition of deforestation. But this figure is also based largely on reports by national forestry departments, which are notoriously unable, and often unwilling, to provide accurate information. The United Nations is preparing a new global forest "census" based on aerial photography and satellite images, which it hopes to publish in the early 1990s.

Where these more objective surveying tools have been used, even by embarrassed countries, the results have tended to show alarming acceleration in losses. Brazil's own National Remote Sensing Program found that the total deforested area of the Amazonian state of Rondonia had almost tripled from 10,000 square kilometres to 27,000 in only three years, from 1982 to 1985.[47] Simple arithmetic suggests that, at present rates of clearing, all tropical moist forests will be gone in about 175 years. But this does not take into account a doubling or tripling of population by that time.

THE HUMAN EFFECTS

The loss of trees outside the great closed forests of the world causes more immediate human suffering than the loss of the jungles.

The open woodlands and "bush" of the tropics provide millions with firewood for cooking and heating, with poles for building houses and making farm tools, with fruits, spices, dyes and fodder for livestock. Some 300 million people in Africa rely on some form of vegetation for cooking, heating and lighting. Some 50 per cent of the population of India and 30 per cent of the people of China do likewise. Globally, three billion people may be consuming wood faster than it is growing by the year 2000.

The very word "deforestation" suggests that something is happening only to forests. But deforestation happens to people. In the hills of Nepal, trees are being cleared not so much for firewood but to make way for farming as populations increase. As trees vanish, women, and to a much lesser extent men, must spend more time collecting wood, and collect less of it. The increased time spent collecting means less time is available for growing food crops. So the family has less to eat and less cash profits from farming to spend on food and other necessities. Women have less time to cook and less wood with which to cook, which means that less is cooked. But older children must also spend more of their time gathering wood and helping out in the fields to make up for their parents' lost labour. Since much of the caring for and feeding of the youngest children is done by older sisters, malnutrition in the youngest increases.

This pattern is repeated throughout large areas of Asia, Africa and Latin America, but Nepal is used as an example because two researchers from the International Food Policy Research Institute spent a year living among the hill people of Western Nepal and studying the direct effects of forest loss.

They found that women were spending about 50–60 per cent more time gathering fuel and other forest products, leaving them 1.4 hours less to work in the fields daily and also less time to cook. They proved that where deforestation was most severe, children were most malnourished, based on standard weight-for-height measurements.[48]

The loss of the rainforests is a global calamity which will eventually affect everyone; the loss of trees and open forests is already a daily calamity for millions of children.

Given rainforest destruction, the loss of trees as growing populations seek farmland, the destruction of Northern forests by acid rain, and the greater threats of global warming to the Northern forests over the next century, then the twenty-first century holds dangers for *all* forests and for growing numbers of people. One UN report somewhat casually lists among the effects of clearing and burning woodland: "water pollution, soil loss and sedimentation, air pollution, degradation of land quality, and loss of valuable habitat for species of birds, animals and plants".[49]

Europe and North America are hardly immune. As acid pollution kills trees in Europe, there will be more wood – from damaged trees – on the market for the coming decade or so. After that, there may be much less. Much of the present damage is in the highlands, and if the cutting of these damaged trees is not carefully planned, then there could be more floods and landslides as water surges down bare slopes.[50] The World Commission on Environment and Development concluded that "Europe may be experiencing an immense change to irreversible acidification, the remedial costs of which are beyond economic reach".[51]

ENDING EVOLUTION

Deforestation provides a dead end for many species and communities of species which have evolved over millennia. The alarm may have been raised too often to bear repeating, but it must be sounded again in any work focusing on future generations.

This may be the worst of all the sorry legacies we are leaving our offspring. "The worst thing that can happen . . . is not energy depletion, economic collapse, limited nuclear war, or conquest by a totalitarian government", said Harvard Professor Edward O. Wilson. "As terrible as these catastrophes would be for us, they can be repaired within a few generations. The one ongoing process in the 1980s that will take millions of years to correct is the loss of genetic and species diversity by the destruction of natural habitats. This is the folly that our descendants are least likely to forgive us."[52]

Rainforests cover only one-sixteenth of the Earth's surface but are thought to house half the Earth's plant and animal species. The total species may number five million, but there may be as many as 30 million; what has yet to be discovered cannot be counted. Of the plant species that have been discovered, only about one in 100 has been studied by science. The larger, more colourful species – the tigers and jaguars and parrots – provide a focus for concern, but it is the more humble species, especially the plants, which may be of more concern to our offspring. We are losing species faster than ever before on the planet.

Evolution occurs more quickly in a hot humid jungle than in a British oak forest; there is more diversity, more competition for nourishment. So jungle plants and animals develop complex forms of germ warfare, complicated chemical compounds to ward off attackers. These compounds far outstrip human chemical ingenuity, and the lists of medicines and industrial compounds derived from them is endless; the hope for future products and medicines is infinite.

We are also losing the genes contained within these species, which could lead to the breeding of new food and medicinal plants, and could make it possible to breed natural pesticides and fungicides into present commercial crops. As scientists become better at genetic engineering, then the need for varieties of natural genes with which to engineer increases rather than decreases. One wry look at the achievements and failures of the 1980s noted that an estimated 100,000 plant and animal species had become extinct over the decade, while only 2,632 new plant and animal species had been patented.[53] It has been estimated that at the current rates of tropical forest loss, humankind may consign one million or more species of plants and animals to oblivion by the turn of the century.[54]

"It really does seem extraordinary that we should be destroying our genetic inheritance at precisely the time when we most need it", said Britain's Prince Charles. "What possible justification can there be for systematically stripping future generations of their options – in a way that defies even conventional logic?"[55]

FOCUSING CONCERN

The destruction of the Amazonian rainforest has recently become a rallying point of global environmental anxiety. That focus may be counter-productive. The thousands of newspaper articles, television programmes, conferences, demonstrations, and expressions of "grave concern" by political leaders – not to mention trips into the rainforests by pop stars – give the impression that progress is being made. In fact, there are few signs of real progress, other than this outpouring of concern.

What can the average citizen do, given that the problem is so far away from home?

In the United States, citizen groups engaged in a little lateral thinking. They noted that much of the destruction of the Brazilian rainforest was being subsidized by loans from the World Bank. But the Bank has traditionally been a closed and secret organization; directly answerable to no single electorate, it is hard to lobby and influence. So US citizens' groups lobbied the most powerful body directly answerable to US citizens: the Congress, which also controls a large slice of the Bank's funding. They alarmed key members of Congress about the fact that US tax money was being used to clear Brazilian forest, and Congress passed that alarm along to the bank.

Bruce Rich, a leader of this effort, is fond of describing the reasoning behind the lobbying strategy: "If you want to influence an organization, give it money. If you want to control it absolutely, threaten to withdraw its funding." Congress threatened to withdraw some Bank funds. The Bank then withdrew funding from some development schemes in the Amazon.

Feeling the pressure, Brazilian president José Sarney announced plans in 1989 to take better care of the one-third of the world's rainforests which lie in his country. These plans included a $100 million, five-year programme to zone the rainforest for economic activities, temporary suspension of raw-timber exports and of tax incentives for clearing forests for cattle ranches, regulations on the production and sale of toxic chemicals used in mining and agriculture, and the setting aside of almost three million hectares of new parkland. He even promised to look into setting aside more land for the forest tribes. It was progress, but the lack of political will behind it was shown when his government reneged on its promise to get the gold miners, and the mercury they are putting into rivers, out of the forest.

British environmental groups, though they have mounted

some excellent information campaigns, have been less successful in alarming their government. Part of the problem is, as usual, secrecy. The government refuses to tell its citizens how the British Executive Director at the World Bank votes on loan proposals. Challenged on this by the citizen's group Survival International, defenders of the rights of tribal people, staff at Britain's Overseas Development Administration (ODA) said that they could not reveal votes on projects because this was confidential to the Bank. ODA staff were then at a loss to explain how the US government can regularly publish its voting position on every single project funded by every development bank, including the World Bank.

When a former British World Bank director was asked in a television interview if he thought that the British people had the right to know how their government was voting on the Bank's spending of British tax money, he said that indeed they should have that right.[56] Not only does the government think its citizens have no right to know the British vote, it also refuses to tell members of parliament.

National Executive Directors are usually advised on how to vote by national aid agencies, and this advice is based on Bank "staff appraisal reports" on the various loan projects. Neither ordinary British citizens nor environmental and tribal rights groups can see these reports. A tribe in Brazil living in rainforest about to be flooded by a Bank-backed dam cannot see them. But Northern businessmen can. They are available to commerce at the Department of Trade and Industry library in London and at the Department of Commerce reading room in Washington.[57] This fact shows to whom the Bank, and the Northern governments which fund it, feel most accountable.

THE CITIES: WORST OF BOTH WORLDS?

One of the most amazing and under-reported stories on the planet today is the rapid growth of Third World cities. It is under-reported by the media largely because it is so complicated, and often falls outside the scope of "environmental" journalists' brief.

It is hard to generalize about this urban growth. Part of the swelling of greater São Paulo, Brazil, now populated by some 16 million people, has been caused by waves of farmers fleeing drought-prone and desertified North-east Brazil. The desertification of North-east Brazil has been caused by government policies. Two million farming families own no land, while an area the size of France is unfarmed or farmed very little because it is owned by the biggest land-owners.[58] Redistribution of land would have kept people in the countryside and limited the growth of São Paulo, the population of which has almost doubled since 1970.

The population of the much smaller city of Nouakchott, Mauritania, increased from 5,800 in 1965 to 250,000 in 1982, almost all of this growth caused by rural dwellers abandoning the parched countryside for the town. Flight from over-used countryside is filling these and other cities, but many more are growing rapidly as a result of the natural population increase of a large, young, urban mass.

The urban population of the Northern nations is expected to double from 447 million in 1950 to 950 million in the year 2000, while the urban population of the Third World is expected to increase almost seven-fold from 287 million in 1950 to 1.9 billion in the year 2000.[59]

Large proportions of these Third World city dwellers, the majority in many cities, are poor people, living either in crowded slums or in shanties they have built themselves, often illegally and often on land illegally occupied or illegally subdivided.

In one sense these poor people living outside the law are the rulers of the cities; in another sense they have no power at all. They rule, because they chose where to "squat", where to erect their shanties and new settlements. The choice is by default. They tend to choose land no one else wants, in hopes of increasing their chances of not being evicted. So they choose the unhealthiest, most dangerous land: marshy ground (Guayaquil, Ecuador),

Opposite Children help to bail out their shantytown home near the beach in Port au Prince, Haiti, after a high tide flooded it.
Mark Edwards/Still Pictures

desert fringes (Lima, Peru), steep hillsides and the sides of ravines (Rio de Janeiro; Guatemala City), land near polluting or dangerous factories (Bhopal, India), or near the seaside, erecting houses that fill with water during high tides (Port-au-Prince, Haiti).

This is the worst land not only because it is the most dangerous and least healthy, but because it is the worst in terms of building or rationally governing a city. The growth often occurs precisely where it is hardest to reach the new homes with roads, sewers, schools, health care and all other amenities. This is the sense in which the poor have no power; they often do not officially exist in the eyes of the city administration, except perhaps as a problem, and a community to be evicted when necessary.

Urbanization is death to children. Some of the reasons are covered in other chapters of this book, such as the high rates of infectious diseases and pollution. But the phenomenon of Third World urbanization deserves at least cursory treatment in its own right, because it is where the worst of two worlds – the world of underdevelopment and poverty and the world of rapid industrialization – meet and combine the damages of both.

CHILDREN IN CITIES

During the reign of Queen Victoria, the overall death rate among the British population declined from 22.2 deaths per 1,000 in 1851–60 to 18.2 per 1,000 in 1891–1900. This improvement must have had to do with improved diets, sanitation, water supply and so on. However, the *infant* death rate was actually higher in the closing years of Victoria's reign than it had been in the beginning, accounting for about a quarter of all deaths in the nation. In 1907–10, in the borough of Bermondsey in south-east London, 135 infants out of every 1,000 born died in their first year, a rate equal to the infant death rate of Bangladesh today.

Anthony Wohl, who provides these figures, also offers an explanation: "As the century progressed, more and more children – and certainly a higher percentage of the total number of children in the nation – were being born in towns. In the early 1890s, when the infant death rate for England and Wales as a whole was around 153, the 28 largest towns had an average infant death rate of 167 . . ."[60]

Today rapid urbanization in the developing world seems to have played a part in increasing the risk of children's early deaths. When Europeans were moving to cities there was rapid industrialization, economies were buoyant, and they were moving to opportunities for employment. In most Third World cities today, there are far more people competing for fewer jobs in economies which are stagnant or declining. A child born in a Third World squatter settlement lacking safe water and sanitation is 40–50 times more likely to die before reaching the age of five than one born at the same moment in a wealthy nation.[61]

Crowding compresses and multiplies the effects of the obvious problems. Water may be hard to get in many rural areas. But in many cities, sewage flows untreated into the rivers; and the poor get their drinking water directly from those rivers. Overcrowding provides a perfect environment for the spread of communicable diseases such as measles, flu, tuberculosis, assorted respiratory diseases and meningitis. Big cities tend to contain most of a poor nation's hospitals, doctors and health workers, but it is rarely possible for the poor to use them. Thus, though people of the Haitian countryside are among the poorest in the world, death rates among infants in the capital of Port-au-Prince are three times higher than the rates in the countryside.[62]

Merchants and markets fill cities with a great quantity and diversity of food. When drought and famine strike a nation, it is largely the poor in the countryside who die. But in normal times, food is widely available in the countryside; most families can grow it, gather it, trade goods for it, and rely on extended families to help out. They can also buy it, usually fairly cheaply. In cities, in normal times, the poor lack most of these options. Thus poor urban people often have a harder time getting enough food than their rural counterparts, and there are often more severely malnourished children in poor groups in urban areas than among poor groups outside cities.[63]

These problems are compounded by the fact that when families move to cities it is often only the mother, father and children who move, just as when the British were moving rapidly into cities. So care-taking and experienced older relatives such as grandmothers are not available to provide support. If both parents work, and they must if it is at all possible, then the children are left on their own in a dangerous, unhealthy environment. Emotional insecurity may be as big a threat to the welfare of urban children as any of the more obvious environmental threats.

GOVERNING CITIES

The political bodies given responsibility for these urban ills – city governments or city councils – are often systematically starved of the political or financial power to do anything about them.

This may take an extreme form when local governments are controlled by opposition political parties. They may have their powers and budgets reduced, or they may be replaced by central government commissioners or appointed officials, the latter often happening after a coup.[64]

But even in normal times, local governments receive only a tiny percentage of tax revenues. The more talented civil servants want to work for the national governments, where the action is. And local officials are generally more easily bribed, or bribed for less money, by commercial interests than are their national counterparts.

Across the Third World, countless groups of the urban poor are forming their own organizations to improve their lots. These often start in response to one specific need – the digging of a drainage canal or the establishment of a communal kitchen to provide cheap food. Such groups often expand their activities into all areas of community life, including petitioning the government for improvements. But most urban needs cannot be met solely by local groups; these include the need for safe water, roads, sewerage, electricity, and systematic refuse removal and health care.

Ideally, the best results occur where citizens' groups team up with the government to work together on meeting these needs. And there are countless examples of such teamwork. But such co-operation takes a great leap of political courage on the part of governments, especially where they have to recognize people who do not officially exist. The urban poor not only often live in illegally built structures on illegally occupied or subdivided land: they may tap illegally into existing electricity or water systems, drink in illegal bars, work in unlicensed businesses and ride in unlicensed buses or taxis. Most governments feel that to recognize their existence is to approve of their illegal homes and livelihoods.

So the only form of official recognition reserved for many of the urban poor is eviction. Millions have been evicted, some several times, over the past few decades in Venezuela, Argentina, Chile, Nigeria, Senegal, Tanzania, South Korea, the Philippines, Thailand and India. In some cases, the shanties are simply bulldozed and the poor are left to their own devices; in some cases they are offered the right to apply for "public housing" – which they cannot afford. In some cases they are moved to other land, often far from jobs, friends, relatives, schools and any coping mechanisms they had devised.[65]

If there is anything worse for a child than being stuck in a dangerous, unhealthy environment, it is being forced to move suddenly and often from one such environment to another.

The answer to these problems within cities lies along the difficult path of partnerships between the poorest citizens and their governments. The city officials will have to forget the latest concepts of urban planning being taught at Harvard University or the University of London, and harness – rather than repress – the energies of the poorest and of their organizations. Governments will never be able to resolve their urban problems without healthy economies, but there is much they could do to relieve suffering without money.

In most Southern cities, tremendous amounts of land are owned by railroads, the military and the dominant church. Such land could be made available for housing the poor near their jobs. The rich who buy land and leave it unused in hopes that it will increase in value could be taxed heavily. Such steps do not require cash, but they do require courage.

Governments could decentralize political power and resources. They could charge the richer city dwellers the full costs of water, electricity and other "infrastructure". And they could also make the big polluters pay. Many of the problems of Third World cities stem from the fact that the rich are essentially subsidized by cheap services to live in them and build their factories there, and then charged little or nothing for the damage those factories do.

In the industrialized nations, there are hundreds of thriving cities outside the capitals. In fact, in nations such as Canada, the United States, West Germany and Australia, the capitals are relatively small administrative centres. People and jobs are elsewhere.

But in most Third World nations, government policies guarantee that the capitals and the biggest cities will continue to grow rapidly. All roads and railroads lead to them; most schools, hospitals, companies, factories and jobs lie in them. A businessman who does not live in Nairobi or Lima or Bangkok may find that getting an import licence may take weeks rather than days or even hours. It is for such reasons that nations like Tanzania, Nigeria and Brazil have gone to tremendous expense to set up capitals far from the former capitals and major cities. China in the mid-1980s tried systematically to create job opportunities and small industries in the countryside to keep people from flooding the already huge metropolises. It is not clear how this policy will fare under the present nervous and repressive regime.

Little foreign aid has gone to Southern cities. One reason is that government aid usually must go to central governments, and these governments do not usually want to spend it on the urban poor or the city in general.

David Satterthwaite of the International Institute for Environment and Development estimates that less than two per cent of official foreign aid goes to improve housing for poorer city dwellers. If all the families who had benefited from aid to housing projects over the past 20 years were added together, they would probably include no more than three per cent of the entire Third World's urban population.

5
MEDICINE, POPULATION AND POWER

> Medicine is a social science, and politics are nothing else but medicine on a larger scale.
>
> *Rudolph Virchow, nineteenth-century cell scientist*
>
> Formerly, when religion was strong and science weak, men mistook magic for medicine; now, when science is strong and religion weak, men mistake medicine for magic.
>
> *Thomas Szasz*

Health Worker: "And you haven't even given him his shots! Why don't you get your children their injections? If they had them, they would be out of danger. The polio boosters and all."
Father: "I did, once."
Health Worker: "You parents are so careless. A mother ought to think about the child's environment. At least if you give them shots, you can save them from the danger of disease."

The scene is a shantytown on the outskirts of Delhi. A local health worker is moving quickly among many mothers, fathers and children, examining the children and chiding the parents to take better care of them. She is also cajoling, threatening, advising and encouraging. She knows they have not the means to follow some of her advice, but has not got time to discuss politics.

Health Worker: "Sister, this child of yours, Salima, how old is she?"
Mother: "Six."
Health Worker: "Six! And she weighs 12 kilograms at six!? That is very frail. What do you feed her?"

Mother: "Rice."
Health Worker: "Give her less rice. Feed her what you cook: potatoes, spinach, vegetables. That's what you should feed a little girl; that's what makes a body stronger. And give her her shots. . . ." (Turning to another woman) "Sister, are you still giving this baby only your own milk? And where do you get your water from?"
Mother: "From the pump."
Health Worker: "From the pump!? He's got diarrhoea and you give him water from the pump? You don't get clean water?"
Husband: "There is the pump here, and it doesn't give clean water. Where do you expect us to get it from?"
Health Worker: "And sister, your child?"
Mother: "He's got a fever. He's been coughing for a fortnight, and has fever since last night. I get him some medicine [antibiotics] and he's fine for a day after taking it, and then gets ill again."
Health Worker: "The children are so sickly here. . . ."
Mother (interrupting): "Children, adults, everyone. . . ."
Another mother: "We queue for hours at the hospital, and then after hanging around there are sent away without medicine. 'Why don't you get private treatment?' they say. Private treatment?! How can we?"

The health worker's name is Manju; in the shantytowns, there are rarely last names. She lives among the families she advises, and is one of them, not a doctor. So she cannot prescribe medicine or give jabs. She hands out vitamins and iron supplements and advises on oral rehydration therapy (ORT). What little training she has came from doctors working with the Saurab Education Society, a private charity which arranges for volunteer doctors to hold a local clinic in a thatched hut every Saturday.

Manju, only 21, has the job of locating the children and parents most in need of a doctor, and of mobilizing the neighbourhood to

Opposite Manju, a shantytown health worker, weighs a baby along the Yamuna River near Delhi.
Meera Dewan/Central Independent Television PLC

get their children to the immunization camps held every few months along the river. There, children are vaccinated against the major childhood killers. She must make her rounds quickly, because she is responsible for some 380 families in the shanties along the Yamuna River. Many of the families are large, and many of the children ill and underweight.

But in just a few minutes talking with anxious parents, Manju is assaulted by all the health problems of the neighbourhood, indeed of much of the Third World. There is little food and little safe water. But there is also little health care; Manju's advice and vitamins are the first line of defence for over 1,000 people. She tells them to get clean water, when there is none, and to cook vegetables which they may not be able to afford. She tells them to go to the hospital for antibiotics, when she knows they may not even be seen; and if they are seen, they may be told to buy medicine from local shops, with money they do not have.

MEDICINE AND MICROBES

It is only fairly recently in human history that medicine and the practice of medicine began to make much difference to the health of many people.

Today, medicine in the form of cheap immunization and through cheap antibiotics saves millions of lives yearly, particularly those of Third World children. Yet the technology of medicine will remain an unreliable saviour as long as people do not have political and economic power over both the threats to their health and threats to their livelihoods in general.

This is particularly true as regards the "population crisis". The delivery of pills and coils and condoms does no good until parents, aided by their larger societies, have the power to keep their children alive and to offer them a sound future. No nation this century has reduced population growth without first having reduced infant deaths. The goal of people-led "primary health care", seen by all governments as the way towards global health in the late 1970s, has remained a mirage. And "health" remains a commodity delivered to people – or more often not delivered – from the outside.

The UN's goal of "Health for All" by the year 2000 will remain a hollow fantasy unless the 1990s witness a series of political rather than scientific revolutions. Little of what brings disease and death to children can be found in medical dictionaries: bad water, lack of sanitation, bad housing, desertification, deforestation, and crowded cities. Few of the cures can be found in hospitals.

As the German doctor and liberal reformer Rudolph Virchow wrote prophetically in the middle of the last century before the advent of antibiotics and most forms of vaccination: "the improvement of medicine would eventually prolong human life . . . [but] improvement of social conditions would achieve its result more rapidly and successfully".[1]

THE ENVIRONMENT OF ILLNESS

Populations began to increase in Europe in the eighteenth century, well before medicine had many cures to offer. Many explanations have been put forward for this: a decline in the great plagues, improved climate, improved agriculture and new foods from the Americas.

But why bubonic plague stopped regularly visiting Europe in the seventeenth century is not clear. Hugh Thomas, in his *An Unfinished History of the World*, writes of this mystery: "The eclipse of the plague was not secured by doctors, not by an international medical programme. It died." He notes that its death coincided with an improvement in climate which brought a new optimism to Europe. And perhaps the improved climate and the new optimism had as much to do with eighteenth century population increases as improved health.[2]

The decline of other diseases in Europe and North America – from tuberculosis to cholera – are more obviously associated with improvements in diet, housing, sanitation and water supplies. All of this is meant to make two points.

First, much of Africa's underdevelopment is due to the fact that it tried to skip directly from peasant, subsistence agriculture to industrialization. No other area of the world – not North America, nor Europe, nor Imperial Russia – managed to develop without first developing agriculture and the people involved in agriculture.

Africa, and much of the developing world, has attempted the same improbable leap in health: investing most of its health budgets in doctors and hospitals, skipping the step of improved water supplies, sanitation and diets. It does not work.

The second point is that the "deaths" of many diseases in Europe – plague, leprosy, malaria, typhus – were as much environmental as medical. Doctors once focused almost entirely on how microbes operated inside the human body. A growing number of biologists have been concentrating on how microbes function in the wider environment.

They study the population dynamics and "habits" of microbes in much the same ways that other biologists study the population dynamics and habits of elephants or whales. It was the realization that the virus which causes smallpox had no animal hosts and no human carriers (people who might have carried the virus without developing the disease) which convinced the United Nations that the disease could be eliminated from the face of the earth, a victory achieved in the 1970s. Similar studies of the polio virus offer hope for an elimination of that disease. From this viewpoint, the study of microbes as natural, living species, all diseases become environmental, and the management of their environments becomes a large part of any successful coping strategy.

VACCINATIONS

Vaccination for smallpox was the only strictly medical practice which had any real impact on death rates before the twentieth century.

Today vaccinations are accepted as a matter of routine. But compulsory vaccinations against smallpox, instituted in Britain in 1853, were a frightening and controversial procedure. Many argued that it was a violation of human liberties, that the state had no right to intervene in something so intimate as personal health, and an Anti-Compulsory Vaccination League was formed and campaigned nationwide. Nevertheless, compulsory vaccination became a key precedent for the now widespread agreement that governments have a right and a responsibility to intervene in the health of their citizens. Today vaccination is possible for such diseases as tuberculosis, polio, diphtheria, whooping cough, typhoid, tetanus, measles, anthrax, rabies, cholera, hay fever, influenza, rubella (German measles) and plague, some of which are more effective and long-lasting than others. It is largely through immunization that modern medicine is having its biggest impact on lives and life expectancy in the developing world.

More than 1.5 million children under the age of five die in the Third World every year of measles (with some estimates as high as two million). Many of those who survive a bout are left prone to malnutrition and other more serious diseases, such as pneumonia. The normally used Schwarz vaccine is not recommended for children younger than nine months. However, many Third World cities suffer measles epidemics as families with unvaccinated children arrive from the countryside. Thus many children die of measles before they can be vaccinated. New vaccines which can be given at four months, and thus increase protection, are undergoing clinical trials.[3]

More than three-quarters of a million children die annually of neonatal tetanus (along with a great many mothers); this is picked up during unsanitary births when the umbilical cord is exposed to spores of the tetanus bacillus. Another half a million children die of whooping cough every year.

But these atrocious figures represent great progress. When the World Health Organization (WHO) began its Expanded Programme on Immunization in 1974, less than five per cent of the children in the Third World had been vaccinated against the programme's six focal diseases: measles, tetanus, whooping cough, diphtheria, tuberculosis and polio. At the time WHO began its efforts, five million children were dying each year of these diseases. Today, about two-thirds of all Third World children have been vaccinated, an effort believed to be saving two million lives a year. But approximately three million are still dying because they have not had their jabs.

Three years after WHO began its work, now supported by UNICEF and other UN agencies, it set the goal of "universal childhood immunization" by the end of the year 1990. This does not mean that every child gets a jab, but that enough children are immunized to stop the transmission of the diseases. This target varies with diseases, but generally requires that at least 80 per cent of the population is vaccinated.

Many countries will not reach the goal. But many, among them the poorest, have already reached or almost reached it. These include Botswana, Cuba, Egypt, The Gambia, Iraq, Jordan, Oman, Rwanda, Tanzania and Saudi Arabia. A large part of the fall in the child death rates stems from the fact that China, with one-sixth of the world's children, is thought to be vaccinating 96 per cent of its children.[4]

This progress has encouraged the UN system to set some other ambitious goals for the year 2000: the eradication of polio; the elimination of neonatal tetanus; a reduction in measles deaths by 95 per cent and measles cases by 90 per cent compared to 1980 levels; and the immunization of 80 per cent of all children under one year of age against diphtheria, whooping cough, typhoid, tuberculosis, measles and polio.[5]

Technically, this is easy. Financially, the cost is low; the total package of vaccines costs less than $1.50 per child. The question, as ever, is whether politicians can find the will – nation by nation – to see these goals realized.

POLIO: THE DEATH OF A DISEASE

Talking of "universal" immunization efforts makes it seem that everything is in the hands of the United Nations. A closer look at the fight against polio shows how all of these campaigns rely on alliances among international agencies, governments, aid agencies, charities, local health workers and the parents getting their children to the health centres or the health workers.

"It is a gift from this generation to the next", said one health worker, speaking of efforts to rid the planet of the virus which causes polio.

The virus, which comes in three types, usually enters the mouth and multiplies in the throat and the intestine. From the intestine, it can penetrate the spinal cord, affect the nerve cells responsible for stimulating the contraction of muscles, and cause various types of paralysis. Once it affects the central nervous system, there is no cure.

It usually strikes children. Only one in every 100 children infected is paralysed, and most of the others have no symptoms. So apparently healthy children can be carriers of the virus and can pass it on. The virus is often transmitted from the faeces of one child to the hands of others and into their mouths. But it can also be carried on contaminated clothing, sheets or household objects and on the bodies of flies. In that it is a disease of unsanitary, crowded conditions, it is a disease of poverty.

It is not a new disease. Egyptian wall reliefs 3,500 years old show a man with a crutch and one thin leg, the foot contorted in the traditional manner of polio. In more recent history, it struck in all nations, including the richest. During the 1950s, 30,000 to 50,000 cases were being reported annually in the United States; and cinemas, swimming pools and other gathering places for children were often closed during the summer "polio season".

Today, about 10 million people are to some extent disabled by polio, with an estimated 200,000 new cases each year out of the 40 million children thought to be infected by the virus. It tends to cripple rather than kill, usually affecting the legs. In the worst affected countries of Africa and Asia, surveys of schoolchildren find four to eight lame children in every 1,000.[6]

The American Jonas Salk developed the first vaccine in the early 1950s, growing the viruses on the tissues of monkeys and then inactivating, or "killing" them. Injected into the bloodstream, the vaccine causes the body to develop antibodies against the polio virus. This was followed by the development of the "live" oral vaccine by Albert Sabin. The viruses in the Sabin vaccine have been treated, so they cannot cause the disease. But they can multiply in the intestine, like the wild virus, and produce immunity there, where the wild virus first strikes.

There has been heated, and sometimes rancorous, debate over which vaccine is the best. Both have their uses. The oral vaccine produces antibodies in the blood, but also local antibodies continuously secreted in the gut. So children immunized in this way do not pass wild viruses through their system to infect unimmunized children. The injected vaccine, even the modern "enhanced" version, requires repeated shots. But it can be mixed with vaccines for diphtheria, whooping cough and tetanus; so one shot protects against four diseases.

The fact that polio affects only humans, and not animals, means that it can be banished from the planet. Already the wild polio virus is extinct over large areas, including Canada, the United States, the Scandinavian countries and Japan; and in 1988, there were no new cases in 25 of the 32 European nations. However, Latin America is actually making progress faster than Europe. The Americas are expected to be free of polio in the first years of the 1990s, with Europe expecting to exorcise the virus by 1995.

The hunting of the polio virus, one of the smallest of all viruses, is a mix of high-tech laboratory work and a lot of personal drama. Health workers carrying boxes of chilled vaccine must often brave raging rivers in dug-out canoes to reach remote villages, slog on foot through monsoon rains and catch rides on tractors or camels from one African village to the next.

Smallpox was easier, in that the freeze-dried vaccine did not need to be refrigerated. Also, there were no carriers. People infected by the virus got the disease, and in about two weeks of infection developed the characteristic deep rash which left most survivors scarred for life. Laboratory testing was not needed; victims could be easily found and all contacts vaccinated. So the disease could be contained by the vaccination of a smaller proportion of the population.

In the Americas, which began their all-out assault on polio in 1985, all cases of paralysis of children younger than 15 are reported; a health worker visits, trying to get there within 48 hours and take a stool sample. Some 6,000 stool samples were examined in 1988 in Latin America. Vaccination in the area of a probable case is begun immediately. Specially-equipped laboratories need one or two months to confirm diagnosis, but modern equipment allows the labs to say whether the virus is local or foreign – that is, brought to the area by an outsider. The very few cases in which the disease is actually caused by vaccination can also be identified.

This effort in Latin America, which expected only 100 new cases in 1989, is expected to cost $520 million over 1987–91, 20 per cent of this donated by international agencies, Northern aid agencies and Rotary International. Put into perspective, that is much less than the cost of producing three new US B–1B bombers. Donald Henderson, who led WHO's successful battle against smallpox from 1966 to 1977, reckons that the eradication of that disease now saves the world $1 billion every year, since neither vaccination nor treatment of smallpox victims are necessary any more. That $1 billion is more than three times the cost of the entire smallpox eradication programme.[7]

Dr T. Jacob John, a professor of medicine in India, reckons that savings could be higher from getting rid of polio. But an estimated two million more children may be crippled by the disease before it is eradicated.[8]

Many of the lame children crawling the streets of Third World cities, begging or selling small items, are polio victims. The other major health effort needed as regards polio is care for victims, often children in the poorest families. Mothers and fathers can be taught simple exercises for their children, so that affected limbs are kept moving. When children sit in one position for hours each day, limbs can become deformed and shrink. Braces must be made and used properly, and changed as children grow. Many countries have begun "community-based rehabilitation" programmes to send rehabilitation workers into homes. Surgery can help some crippled children to walk, but this is not an option for the majority of the world's polio victims.

ANTIBIOTICS: TOO MUCH AND TOO LITTLE

The saga of antibiotics should encourage the conservation of all species, in that it shows how even the most humble living organism may have a profound effect on the fortunes of humankind.

The Scottish physician Alexander Fleming noticed in 1928 that a mould, which apparently drifted in through an open window, was able to destroy the bacteria with which it came in contact. This led to the development of penicillin in 1940. Later, Dr Selman Waksman in the United States discovered the basis of streptomycin and other antibiotics by routinely studying micro-organisms in the soil.

The word *antibiotic* comes from the word *antibiosis*: the opposite of symbiosis, when two organisms co-operate to their mutual advantage. In antibiosis, one organism has a destructive effect upon another; penicillin and all of its derivatives are based on natural products created by one micro-organism to destroy another.

Despite the millions of lives they have saved and continue to save each year, antibiotics are now viewed with grave suspicion in the North because they are widely over-used. Doctors casually prescribe them for viral ailments such as colds and other illnesses against which they are not effective.

The more bacteria are exposed to antibiotics, the more strains of bacteria become resistant to them. The number of intestinal bacteria showing signs of resistance tripled between 1976 and 1981, and this has led to fears of a resurgence in such infectious diseases as pneumonia, meningitis, cholera, salmonella and venereal diseases.[9]

Widespread use of antibiotics in beef cattle and other livestock also causes rising resistance. "Every animal or person taking an antibiotic becomes a factory producing resistant strains [of bacteria]", the *New England Journal of Medicine* reported.[10] "And fears have been raised that at current levels of over-use, we can look forward to a period in which 80–90 per cent of the infectious strains which arise are resistant."[11] An Australian report described the growing resistance as an "environmental hazard", as it changes the microbial environment which spawns diseases.

ANTIBIOTICS IN THE SOUTH

Antibiotics are also over-used and misused in the Third World; but the problems there are, as usual, different.

In many countries, antibiotics are available without prescription in the marketplace. The WHO list of 200 essential drugs contains only 16 antibiotics; there are 200 on the market in Central America. In the South, they still have much more of a "wonder drug" reputation than they do in the North, so people will pay a great deal of money for these alleged cure-alls. And they use them in an attempt to cure everything, including illnesses and injuries upon which they can have no effect.

Even when buying the right antibiotic for the right disease, a customer may be able to afford only two or three pills, not realizing that a whole "course" of doses must be taken for the medicine to be effective. Under-dosing builds up bacterial resistance. It would be tragic if the misuse of these drugs created an environment loaded with super-bacteria, in a part of the world where bacterial diseases are still major killers.

The other side of this problem is the under-use of antibiotics in the Third World. Acute respiratory infections, including pneumonia but excluding measles and whooping cough, kill 2.2 million children under five in the Third World each year. Most of these respond well to oral antibiotics. But most countries forbid health workers other than doctors to give antibiotics, so obviously these workers are not trained in their use. Given the extent to which even highly trained doctors over-use these drugs in the North, there is some wisdom in this approach.

But there is mounting evidence that properly trained village health workers can use antibiotics effectively, and save millions of lives. To tell whether a child apparently suffering from only a cough or cold has something more, and is in life-threatening danger, one need only count the breaths: over 40 per minute in children younger than one and over 50 in older children are signs of risk. In the Punjab as early as 1973 health workers taught to use antibiotics halved the numbers of children's deaths from pneumonia.

Reviewing six recent trials in Asia and Africa, Britain's Dr Felicity Savage reckons that there is now ample evidence to suggest that with regular supervision, village health workers can use antibiotics effectively without abuse or over-use. In a Tanzanian study, village health workers using antibiotics reduced children's deaths from pneumonia by more than 67 per cent within five years.[12] Dr Savage further argues that effective use of these drugs by such workers also increases the acceptability of primary health care programmes and gives villagers more faith in primary health care workers.

Drug companies heavily advertise in the South medicines which treat only the symptoms of respiratory diseases – such as cough medicines and decongestants. These are "expensive, ineffective and sometimes harmful in the treatment of acute respiratory infections", according to Dr Robert Douglas. He adds that they are "often heavily promoted and lead to false expectations. Many parents and some health workers wrongly believe that these medicines can save their children's lives, and because of this some children therefore miss out on the real life-saving treatment".[13] It is the story of baby formula and anti-diarrhoeal drugs all over again.

The waste of lives is enough to make even a UN agency furious. Noting that the efficiency of ORT and oral antibiotics was proved two decades ago, and that 100 million children have since died from conditions both can cure, UNICEF argues:

> It is as if a cure had finally been found for cancer but then little used for twenty years. Diarrhoeal disease and acute respiratory infections kill more people than all the different cancers put together, and most of their victims are not over 50 but under 5. The decisive difference, to be explicit about it, is that the victims of diarrhoeal dehydration and respiratory infections are predominantly the children of the *poor*.[14]

UNICEF goes on to argue that even when one adds together the cost of delivering the three cheap, effective life-savers – ORT, jabs and antibiotics – most of the 14 million lives of under-fives lost every year could be saved for the relatively small additional cost of $2.5 billion per year by the late 1990s.

Should that cost not seem small, UNICEF offers these comparisons: it is two per cent of what the Third World spends on arms yearly; it is the cost of five stealth bombers; it is what US companies spend yearly to advertise cigarettes; it is what the Third World spends every week to service its debts; it is the whole world's military spending, for one day.[15]

POPULATION: THAT IS, BABIES

Another way to ensure that more children survive is to have fewer babies. And a way to ensure that fewer babies are born is to ensure that more children survive.

Rapid population growth is caused by many people having many babies. After saying that, it is hard to discuss "the population crisis" without constructing a tower of clichés and half-truths which collapses under the slightest pressure. There are even those who argue that there is no such thing as a population crisis, that more people mean more consumers and more wealth. While this may be true for parts of the planet, it is not true for the planet as a whole. But those who argue that the population crisis is the chief cause of the environmental crisis are also wrong.

Globally, there is a population crisis. The present world community of over five billion is expected to double in the next century and perhaps almost triple, reaching 14 billion, according to UN figures released in mid-1989. The earth is an almost closed system, receiving from outside nothing but solar energy to provide new energy to support life. Pre-historic sunlight has already laid down deposits of coal, oil and gas – the fossil fuels. These provide most of the world's energy, and once used, they are gone forever.

But present solar energy also keeps us going year after year, providing food crops and fodder, meat, fuel, fibre and wood. One group of experts reckons that the five billion human beings are already appropriating for themselves one-quarter of all the productive power of solar energy that falls on the earth – both land and sea – and 40 per cent of that which falls on the land.[16] It is hard to see how a population of 14 billion would leave much solar energy for other species on earth besides the human. Professor Herman Daly of the World Bank points out the implied end of the world in such calculations:

> Taking the 25 per cent figure for the entire world it is apparent that two more doublings of the human scale will give 100 per cent. Since this would mean zero energy left for all nonhuman and nondomesticated species, and since humans cannot survive without the service of ecosystems, which are made up of other species, it is clear that two more doublings of the human scale is an ecological impossibility, although arithmetically possible. More than two doublings is even arithmetically impossible.[17]

That is the front end, the consumption end, of the problem.

The back end, the pollution end, raises the question of how 14 billion would be able to produce enough primary energy to survive, when the pollution-driven global warming caused largely by the energy production of the existing five billion already threatens the survival of many.

Optimists hope that technology – nuclear fusion, genetic engineering, better solar power – will provide solutions. If we have learned anything about technology, we have learned that it is expensive and becoming steadily more expensive, and that it is usually designed to meet the needs of the rich, who design it. Ninety per cent of the expected population growth in the next century will be in the Third World, most of which cannot even afford electricity, much less nuclear fusion, should fusion ever prove a practical reality.

At the national level, population becomes a more complex and controversial subject. If the size of a population is only a problem when resources are consumed in excess, then the population crisis lies in Europe, North America and Japan. These populations are stable, in that they are increasing only very slowly, if at all. Stable populations are said to be good for the environment. These populations are highly educated. Education is said to be good for the environment.

Yet the 1.2 billion who live in the more developed world are producing three-quarters of the industrially derived carbon dioxide which is warming the planet, and almost all of the chemicals which are attacking the ozone layer. They are also consuming three-quarters of the earth's resources, including three-quarters of the primary energy, while the almost four billion who live in the less developed nations are consuming one-quarter of the planet's resources. The rich world is importing, at going-out-of-business-sale rates, the ecological carrying capacity and hopes for a sustainable future of the poor world. The poor

Opposite A Sudanese nomad mother and child. When children have a greater chance of staying alive, mothers have fewer children. Educating women is the best way to keep children alive and produce smaller, healthier families. *Mark Edwards/Still Pictures*

nations are willing to sell this birthright, as it is their only means of earning hard currency.

Many developing nations have their own individual population crises because their already large populations are growing rapidly, and because these populations are mainly rural and are thus over-using soil fertility and water supplies both to feed themselves and to produce exports.

Africa, though not densely populated, is growing fastest; its population of about 650 million, 70 per cent of whom live in the countryside, will at present rates of reproduction double in only 24 years. The more than three billion Asians, two-thirds of whom are rural, will become six billion in 38 years. Even urbanized Latin America, whose 448 million people will double in 32 years, has about one third of its population living directly off environmental resources in rural areas. What can countries or continents do to slow things down?

First, the people producing the large families must want to have fewer children; and many of them already do. Some 300 million couples in the Third World do not want more babies but have no reliable means of preventing more births.[18] The findings of the World Fertility Survey suggest that half the women of child-bearing age in the Third World do not want more children.[19] If women who did not want to become pregnant could actually exercise their choice, the rate of population growth in the Third World would fall by about 30 per cent.[20]

Despite this need, the simple provision of more pills, coils and condoms is not the best way of helping couples limit families. That approach alone is as irrational as trying to meet people's needs for safe water by providing them with buckets. Hardware is a part of but not the whole answer.

Keeping children alive is, strangely, the best way of guaranteeing fewer children. Even today, many otherwise well-educated Northerners cannot understand how "do-gooders" can worry on the one hand about high population growth rates and on the other about high infant mortality rates. Surely that is a contradiction of concerns? Is not high infant mortality "Nature's way" of keeping population at bay?

The plain fact is that parents will not even consider birth

control until they know that most of the children they do have will live. So more babies must be saved before there will be fewer babies. Previously, there was a long "lag time" – perhaps decades – between falling infant death rates and falling birth rates. In some nations today that lag time has dropped to a few years.

Countless studies proving this led UNICEF Executive Director James Grant to claim flatly in Moscow in early 1990 that "No country has dramatically reduced its fertility rates until it has first significantly reduced its child death rates". Experts would argue that this was not true for Europe in the nineteenth century; but Grant argues it is true for every Third World country after the turn of the century.

PARENTS AND POWER

Francis Moore Lappé is best known as a campaigner on food and hunger issues. But she so often ran up against the myth that "nations are hungry because they are over-populated" that she decided to write a book to shatter the myth. In *Taking Population Seriously* she and her co-author, Rachel Schurman, demonstrate how "the varied forces keeping birth rates high [are] aspects of a systematic denial of essential human rights – understood to include not only political liberties, but access to life-sustaining resources and to educational and economic opportunity".[21]

Lappé and Schurman agree with others who find that a rising educational level among women is the factor most strongly associated with declining birth rates. Other factors include access to land and other ways of getting enough food, increased opportunities for women to work outside the home, and improved health care for women.

Zimbabwe has achieved the highest rates of use of modern contraceptives in all of Africa by establishing a system of women's health and education workers in rural areas. While helping women stay well, while teaching them to read, write and keep accounts, while showing them how to make money from gardening and small business projects, the workers also offer the women contraceptives. And the women accept them, but only after they have been made confident that their future welfare lies in their own hands, rather than in the hands of numerous offspring.

Such efforts show up the old slogan "development is the best contraceptive" for the half-truth that it is. Any old "development", especially development measured only in increased national wealth, will not do. It is the poorest who are having the children, and it is the poorest for whom children have most economic value. Lappé and Schurman amassed studies showing the huge financial contributions children make to a family from a surprisingly early age. In Java, a boy may assume responsibility for the family's poultry by the age of seven. At nine he cares for goats and cattle, and harvests and transplants rice. By 12, he can work for wages. By 15, he has paid off the entire investment his parents have made in him.

In Bangladesh, a son is contributing more than he is costing by the age of 12. Girls are equally important, whether they are working in the home, taking care of brothers and sisters, or earning wages picking tea, as girls as young as 12 do in India. Similar patterns persist throughout the rural areas of Africa and Latin America. But even in cities, among poor families who lack steady jobs, children may earn a high proportion of the family income.

The sort of development which acts as a contraceptive is the sort aimed directly at the poorest. It gives them power over their lives. But poor people who begin to sense opportunities to better themselves are dangerous to many governments, so many governments follow policies aimed at keeping the poor poor.

The starkest and most obvious example is in South Africa, where 15 per cent of the population, the whites, officially oppress the rest, especially the three-quarters of the population who are black. The policy of apartheid forces much of the black population to live in "homeland" areas which are too environmentally degraded to support them. The results include high rates of malnutrition, disease and infant mortality among the blacks, and accelerated environmental destruction of their lands. South Africa is "developed"; it has a per capita Gross National Product of $1,800, almost three times the African average. Yet because a reasonable share in that wealth is denied to the poor majority, it not only has the social and health problems of a poor nation, it has population growth rates higher than such poor African nations as Chad and The Gambia.

But the other side of this coin is encouraging. Nations and even parts of nations can lower their birth rates without achieving high average national incomes. China, Sri Lanka and Cuba all lowered total fertility rates (the average number of children a woman will have over her reproductive lifetime) by more than 35 per cent over 1960–85, Cuba by more than 50 per cent. Much of this is attributed to programmes which guaranteed affordable food for the poorest. China's success has recently slipped, and Sri Lanka has discontinued much of its food security system.

The Indian state of Kerala, with population densities three times the Indian average, achieved a fertility rate reduction of 38.5 per cent over 1965–80, partly by "Fair Price" shops which kept the price of food and other essentials low, but also by improved health care, social security payments, pensions, and unemployment benefits for the poorest, and at least the beginnings of land reform. These reforms also lowered infant mortality rates to one-third the national average.[22]

Thus policies which give the poorest more social security, in the broadest sense of more economic and political power, not only slow population growth rates, but also help to keep children alive. Such policies can also allow the poorest to manage their environments, their soil and crops and trees, more rationally, rather than having to over-use them to survive.

These lessons have put the leaders of the Northern democracies in a difficult position. During its hearings into the African droughts and famines in 1985, the members of a US Congressional sub-committee kept harping on Africa's "population crisis". More recently, Margaret Thatcher singled out population growth as the most important factor in the "environment crisis" in a 1989 speech to the UN General Assembly:

> More than anything, our environment is threatened by the sheer numbers of people and the plants and animals which go with them. When I was born the world's population was some two billion people. My grandson will grow up in a world of

> more than six billion people. Put in its bluntest form: the main threat to our environment is more and more people, and their activities . . .

Political leaders of the North, where populations are stable, love the population issue. They feel comfortable with it. They, whatever else they have done, are not the fathers or mothers of those Third World children.

But if development aimed at the poorest is the best way to lower population growth rates, then any obstacles placed in the way of such development keep growth rates booming. The debt crisis, support to corrupt Third World regimes resistant to a redistribution of land, trade barriers against Southern commodities, aid which supports the wealthier rather than the poorer, the aggressive sale of arms to the Third World, exports of hazardous chemicals and inappropriate drugs: all of these are things controlled in part by decisions in Northern nations; all of these are policies supported by most Northern governments. All hinder sustainable development for the poorest.

Mrs Thatcher did not mention any of these issues in her speech to the United Nations; they would have made her uncomfortable. In fact, she argued defensively, "It is no good squabbling over who is responsible . . ."

MOVING TOWARDS HEALTH?

Like the "population crisis", the health crisis afflicting Third World children is driven by present political policies, South and North.

This fact has led many doctors and health experts to criticize strongly the ways in which the UN system is promoting things like ORT, vaccinations and antibiotics in the Third World. In fact, some criticize such promotion in general. To understand why experts could apparently set themselves against these seemingly cheap, life-saving technologies, one has to drop back more than a decade to a meeting held deep in Soviet Asia in 1978. And to understand the reasons for this historic meeting, one has to look at the deepening world health crisis of the 1970s.

It became obvious in the 1970s that most of the planet's people were getting no regular health care. Worse still, the systems of health care being set up by most of the Third World governments seemed to guarantee that they would get no care in the future.

The Southern nations were copying the health care systems of the North, based on expensive urban hospitals full of expensive equipment and run by doctors whose expensive training copied the training of Northern doctors. "Three-quarters of Third World doctors work in cities, where three-quarters of the health budget is spent. But three-quarters of the people and three-quarters of the ill-health are in the rural areas", according to one survey during the late 1970s.[23]

One new hospital in Lesotho was due to cost 40 per cent more than the entire national health budget. In Ghana, 90 per cent of the rural population were sharing only 15 per cent of the total health budget. In Niger, 95 per cent of the people lived in the countryside, but only nine per cent of the doctors lived there. In India, 80 per cent of the people were rural, along with only 30 per cent of the doctors. England and Wales had one doctor for every 760 people, while Bangladesh had one for over 15,000 and Ethiopia one for almost 70,000.

The diseases which the Third World doctors were being trained to treat – heart and circulation problems, bronchitis and other diseases related to smoking, and all the cancers – were not the diseases of their patients. Their people were sick and dying of combinations of such things as malnutrition, diarrhoeal diseases, malaria and other parasitic diseases.

These facts encouraged many health experts, governments and UN agencies to proclaim a whole new view of health care under the broad title of "Primary Health Care". "Health", according to WHO, was not the absence of disease, but was "a state of complete physical, mental and social well-being". This remains WHO's definition of health today, as it and other UN agencies seek to achieve the goal of "Health for All" by the year 2000.

In 1978, after all areas of the world had been asked to draft their own view of Primary Health Care, representatives from 134 governments and 67 UN organisations gathered in Alma-Ata, USSR, to try to figure out what it all meant and how to move forward. The then heads of UNICEF and WHO, which jointly sponsored the meeting, wrote a visionary paper to inspire delegates.[24] Their vision makes extremely depressing reading in the 1990s; so little of what was agreed at Alma-Ata has been realized.

The leaders of the two organizations noted that four-fifths of the world population had no access to any permanent form of health care, and that even those getting urban health care had become "cases without personalities"; there was too much concentration on "medical technologies". Health systems were too often devised outside the mainstream of social development and restricted to medical care, whereas the sources of much ill-health lay in industrialization and a changing environment.

They offered instead a view of Primary Health Care (now known fondly as "PHC") which focused on nutrition, water, sanitation and family planning, as well as immunization against and treatment of the diseases people really had. Health care should be provided largely by community health workers based within the village, town or slum, using doctors and central hospitals as resources only when necessary. The key element of PHC was that it would not be "delivered" from outside the community, but rather constructed by the community:

> There are many ways in which the community can participate in every stage of primary health care. It must first be involved in the assessment of the situation, the definition of the problems and the setting of priorities. Then, it helps to plan primary health care activities and subsequently it cooperates fully when these activities are carried out.

These brave words have almost nothing to do with the reality of 1990, or with events during the 1980s. Almost as soon as the Alma-Ata declaration was agreed, some experts argued that the ideals were unrealistic. They urged that diseases be targeted by experts, and cost-effective programmes planned by governments and agencies.[25]

Critics today argue that it was "selective" PHC, rather than real PHC, which emerged from the Alma-Ata meeting. Governments, UNICEF and WHO are in fact planning what health care there is outside communities and delivering it. They have skipped the community planning and control ideals and gone right to the last element, in which the community "cooperates fully when those activities are carried out".

US geographer Ben Wisner maintains that this approach is actually harmful to sustainable development because it thwarts any attempts by communities to argue and struggle and determine their own health needs and ways to meet those needs. ORT is packaged in capitals and then advertised over television and radio, the same media that advertise cigarettes and harmful anti-diarrhoea drugs. Vaccines, he says, are often delivered by the same military helicopters that normally terrorize the population.

He claims that instead of building a "safety net" for children and families, the UNICEF-type interventions actually create a "floor" upon which poverty built by government policies can become a permanent social structure. The safety nets in Europe and North America are woven not of health technologies but from minimum wages, unemployment pay, sickness benefits and family allowances.[26]

Susan Rifkin and Gill Walt at the London School of Hygiene and Tropical Medicine raise similar points, noting that selective PHC negates the idea of community participation; gives attention only to those with selected diseases, leaving the rest to suffer; reinforces authoritarian attitudes; and has questionable moral or ethical values in which foreign and elite interests overrule those of the majority of the people. The basis of real PHC, they say, is "not merely health service improvements. It is understanding and improving the range of social, political and economic factors which ultimately influence the improvement of health . . ."[27]

Some of the critics also come from within the international bodies. Dorte Kabell of OECD argues that high-profile, single technology health campaigns in the long term "have been found to fragment recipients' health care systems and render recipient administrators permanently dependent on external inputs and know-how".[28]

Lest some of these arguments sound idealistic and academic, one need only imagine a Britain in which there was no National Health Service, unemployment pay, sickness benefits or family allowances; in which only a third of adults could read (as in Somalia); and in which less than a third of the people had access to safe water (Nepal). What then would be the response of the thoughtful British citizen to co-operation between a British government and a UN agency involved in a programme of vaccination and ORT distribution – so more children could survive to become unhealthy, malnourished, unemployed and impoverished adults?

It is arguable whether any Northern countries meet the ideals of PHC. But the average British citizen does, in theory, have a personal relationship with a local doctor, is visited by health workers and has access to water, sanitation, gas and electricity in ways that at least allow for complaining and protesting. The fierce public debate over what many see as the dismantling of Britain's National Health Service shows that common people have a say, if not a deciding voice.

UNICEF has mounted a spirited defence against its critics, noting first that Alma-Ata could not have foreseen that the 1980s would see the severest economic restraints and cut-backs for half a century. It argues that promoting technologies like ORT "has often substantively helped to accelerate the building of PHC infrastructure". UNICEF never intended to spend more than one-fifth of its total budget on immunization and ORT, and in fact spends more on low-cost water schemes than on immunization and almost as much on education. But it admits that "technologies alone will not make much difference" and calls for a great deal more debate on how PHC can be made to work in the real world.[29]

The ideals of PHC, if they ever had a chance, were killed by events of the 1980s, the "lost decade". The debt crisis, stagnant or falling commodity prices, general economic downturns, and repressive regimes not only dealt a death blow to those ideals, but revealed the fragility of selective PHC, as in some parts of the world health and education budgets have been cut, wages have fallen, malnutrition has increased, and the fall in infant mortality rates has stopped or actually been reversed.

Surprisingly little foreign aid has gone to Third World health problems, particularly PHC, and particularly the broad concerns of environmental health. The OECD recently reviewed the money its member governments give as aid and found that 7.2 per cent goes to health. But only slightly over four per cent goes to PHC, and only a little over one per cent goes to such environmental health measures as water supply, sanitation and sewerage.[30] This is strange, given that the wealthy nations were also at Alma-Ata and were among the most vociferous champions of PHC.

David Satterthwaite of the International Institute for Environment and Development has been monitoring aid spending on the poor, and estimates that during the 1980s "less than nine per cent of commitments of the World Bank went to a combination of water supply, sanitation, health care centres, health education, immunisation, nutrition, family planning, disease control, low-income housing and integrated community development projects". Figures were roughly the same for the Asian Development Bank and the Inter-American Development Bank.

Projects which work with the poor to improve health, housing and education "can deliver substantial benefits at relatively low per capita costs", maintains Satterthwaite. "But such projects are more difficult to implement and evaluate, take more staff time per dollar spent and have higher recurrent costs, so agencies do not like them."

Health care systems relying on packets and jabs organized by outside agencies, whether UN or aid agencies, are fragile things. They last only as long as the fashion and cash for such interventions last. The AIDS epidemic in Africa already threatens to pull more cash into urban hospitals and high-tech medicine, as AIDS so directly threatens young urban adults. There may be in future even less money, fewer health workers, and fewer clinics focusing on the environmental diseases which kill children in their millions.

Increased local pollution and the global pollutants behind the greenhouse effect and ozone depletion, along with increased desertification and deforestation, threaten to disrupt what progress the delivery of technology has brought.

6

PERILOUS FUTURE – GLOBAL WARMING AND ALL THAT

> Environmentalists and politicians can argue the costs and benefits of international action on global warming from now until doomsday, and they probably will.
>
> *Elliot Richardson*

Yasser, eight years old – or perhaps nine, his parents were not sure – was so weak he could hardly hold on to the back of the donkey. He, his parents and his four brothers and sisters had been walking and riding the donkey for six days, and for four of those days Yasser had suffered from diarrhoea.

They had left their home in northern Ethiopia for two reasons: environmental destruction and physical destruction. The land was eroded; there had been no rains; there was no food left. But then when an Ethiopian bomb meant for Eritrean fighters blew up the next-door neighbour's oxen, they decided to flee.

"I woke up in the morning and saw my parents loading the donkey", said Fatima, at nine, or perhaps ten, the eldest child and responsible for taking care of her three younger siblings while her mother, Mariam Hamid, took care of the baby. "The hardest part of the walk was when we had to rest the donkey and all had to walk. We stayed under trees during the heat of the day, and we walked at night. But the water got so hot we could not drink it."

The family did not even have a goatskin for the water, but had used plastic containers. Goatskins keep water cool by slow evaporation through the skin, but the water in the plastic got almost boiling hot, Fatima said.

Ibrahim Ismail, the father, said he had not known where he was going; he was just leaving Ethiopia and making for the Sudan. They joined a few other families, with camels, toward the end of their trek, and all wound up at the huge refugee camp of Wad Sherife just across the border in the Sudan.

Wad Sherife was famous during the height of the 1984–6 African droughts, when it held well over 100,000 refugees, with 3,000 arriving most days. Today, with 50,000 people sitting in a treeless desert and 80–300 more arriving each day, it has been largely forgotten by the outside world. Aid agencies and charities are cutting back on their activities amid the administrative chaos of present day Sudan, and thousands more refugees are expected in the near future. Ibrahim said ten families had already left his village and five more were leaving soon.

Dr Yasser Abdul is the medical officer at Wad Sherife's "screening centre" a few kilometres from the actual camp. Ibrahim's family was to be kept there several days to be vaccinated and checked for disease. Dr Abdul treated young Yasser's diarrhoea, but no one was certain that he would survive. The doctor reckoned that he had received and examined 15,000–20,000 people over the two years he had been at the camp. About half the new arrivals had malaria; there were many cases of measles, pneumonia and tuberculosis. And he treated innumerable scorpion bites.

Once they arrive at the camp itself, families are given a tarpaulin to spread on some sticks for a home and enough food for a month. The food allotment is based on a daily ration – the same for adults and children – of 660 grams of grain, 60 grams of legumes (peas etc.), 30 grams of oil, 10 grams of sugar and five grams of salt. It is enough to keep people alive.

Overleaf The refugee camp at Wad Sherife, just inside the Sudanese border near Ethiopia. If global warming dries out the region further, there will be many more refugees and many more such camps.
Inset Yasser and his little sister upon their arrival at Wad Sherife refugee camp in the Sudan after six days on a donkey.
Bruno Sorrentino/Central Independent Television PLC

Ibrahim had no way of hearing him, but while he and his family were walking, the Sudanese commissioner for refugees said his nation could accept no more refugees. However, they will keep coming, fleeing perhaps the poorest nation on earth for one of the next poorest.

Looking around the desert camp and the family's new sticks-and-sheet hut, a long walk from any water, which she would have to fetch, Fatima did not complain but she did say simply, "I miss my friends". Perhaps her friends will soon join her.

People are on the move all across the Sahel, not as dramatically as out of drought-and-war-torn Ethiopia, but a steady, unstoppable flow southwards out of the parched lands into the richer coastal nations of West Africa. There is occasional fighting and sometimes armies force people back northwards.

This is today. If the scientists are right about global warming, the steady flow of environmental refugees is likely to become a flood, worldwide. Some UN officials are predicting whole nations of "boat people" as lands dry and sea-levels rise. But there may also be more refugees from strife, as local wars develop over water and arable land. Egypt and Ethiopia have no treaty regarding use of the Nile, and Egyptian officials have vowed to go to war to protect their use of that water.

There may be camps like Wad Sherife, full of thousands of Fatimas and Yassers, throughout the poor world.

TOWARDS A DIFFERENT PLANET

Trends in the world today proffer a mixed message. On the one hand the message is of gloom and despair: millions of children die needlessly every year in the poorest countries, largely because their governments cannot and will not invest enough in their well-being and because the richer nations make it difficult or impossible for poor nations to develop.

On the other hand, there is the message of hope: over the long term, infant mortality rates have fallen because some poor governments, some aid agencies of rich governments, and the UN agencies have invested in immunization, education and general health-care programmes.

The progress comes mainly in the form of technical fixes, paid for in large part by outside money and organized to a large extent by outside experts. But, given the low priority most governments place on children's health care and education, these technical fixes cannot maintain progress in the face of such shocks as recession and falling commodity prices.

Global climatic change and ozone depletion may provide far greater shocks than any which have gone before. Climatic change will affect all nations, but will affect each nation differently. It may disrupt the economies and agricultural systems of many rich nations, meaning that they will have less surplus money, food and concern for the Third World.

But it is expected to hit the poorest nations, and the poorest people in those nations, hardest. Global warming and ozone depletion may be relatively new threats, but for many poor people in the developing world, many of the effects will feel like more of the same: low crop yields, malnutrition, disappearing forests, poor health, and water scarcities. The only really new impact, and perhaps the most frightening when children are considered, is the fear that ozone depletion may disrupt people's natural immune systems. Thus vaccination programmes could turn from life-saving to life-threatening exercises.

THE NEW SHOCKS: THE BASICS

Most people have at least a vague idea of what global warming and ozone depletion are about, but the notions can be alarmingly vague. A recent conference of European teachers found that some thought the planet was warming because more people had central heating; others thought the heat was caused by sunlight pouring through the "ozone hole". It would be nice if things were so simple.

Human activities and many natural events release into the atmosphere "greenhouse gases", which are transparent to short-wave solar radiation. The sun's rays pass through these to warm the earth. This heat produces longer wave radiation. An unpolluted atmosphere allows much of this long-wave radiation to pass through into space. But the greenhouse gases are opaque to the long-wave radiation and do not let it through, so the planet warms.

The proportion of these gases amid all the atmospheric gases is and will continue to be tiny, measured in parts per million of the total atmosphere. The most important gas is carbon dioxide (CO_2), released by the burning of fuels and burning of forests and grasslands. Decomposing wood also releases carbon.

Another greenhouse gas is methane (swamp gas), released by the digestive systems of cattle and other grazing animals, by termites and by swamps – natural and unnatural. The number of grazing animals continues to increase steadily; so does the number of termites, as forests are cleared; so does the amount of land covered in rice paddies, an unnatural swamp which releases methane. Nitrous oxide is another greenhouse gas; it is released by the decomposition of organic matter and by burning fuels, forests and grasslands.

CO_2 is thought to be responsible for half the warming, and all the other gases for the other half. So there is no single villain in terms of chemicals or of human activities. Nor is there one villain in terms of nations or regions of the world. About three-quarters of all human-released CO_2 comes from the North, but the South's share is rising as it industrializes and cuts more forests.

It is hard to get two scientists to agree on even the simplest things, such as where to eat lunch or how to run a committee. Yet over the past few years, scientists have been gathering in their hundreds from all nations of the globe, and *agreeing* on roughly what will happen when all these gases reach levels more or less equivalent to a doubling of the CO_2 atmospheric content which existed before the Industrial Revolution, something that is expected to happen around 2025–35. The agreement is based not so much on the small warming – a fraction of a degree – which is already thought to have occurred this century, but on computer models of how the global climate works.

The models predict that a doubling will warm the globe between 1.5 and 5.5°C, the warming increasing the further one gets from the equator. It will also raise sea-levels between 20 centimetres and 1.65 metres, most of this from an expansion of water as it heats, with little effect yet from melting ice.[1] A rise of

80 centimetres should be more than enough to flood unprotected coastal lands.

This does not mean that these changes in temperature and sea-levels will have occurred by 2025–35, but that the doubling will cause these changes eventually; there will be a lag time. Richard Warrick and his colleagues at the University of East Anglia predict a warming of one or two degrees centigrade by the year 2030, assuming that present rates of pollution continue.

This warming in such a short time would be a tremendous shock to the planet's system and to human civilization. The coldest the planet has been, during the Ice Ages, was only five degrees colder than it is now, and a five-degree rise would make the planet warmer than it has been for two million years. And five degrees is no magic upper limit; it is only the upper limit of the effects of doubling. If humans go on indefinitely increasing the production of greenhouse gases, the globe is expected to go on getting hotter – indefinitely.

There is tremendous uncertainty about all of this, and will continue to be until it begins to happen – or does not happen. The only way to prove that it will happen would be to find another planet exactly like the earth, raise greenhouse gas concentrations there and watch the results. At the moment, we are conducting the experiment only on earth. A few scientists do not think there will be a warming. Others argue about feedback effects. A warmer world would mean more evaporation, so more clouds, and these clouds might cool the globe. But there are other possible feedback effects in the opposite direction. A slight warming might release much of the methane and CO_2 thought to be trapped in the frozen Arctic tundra, accelerating the expected warming. The arguments go on.

Chlorofluorocarbons (CFCs), used in some aerosol cans, as coolants in refrigerators and as a foaming agent in plastics, are also greenhouse gases. But they have another effect, the destruction of the ozone layer. As CFCs rise into the stratosphere they slowly break up, releasing chlorine and other chemicals. These react with the high-altitude ozone and destroy individual ozone molecules, which are formed by three oxygen atoms. But the chemical reaction does not destroy the chlorine, which may remain in the atmosphere destroying ozone for as long as a century.

In fact, most of the greenhouse gases have a long average "resident time" in the atmosphere: CO_2, 100 years; nitrous oxide, also an ozone destroyer, 170 years; methane, 10 years. So we have already committed the planet to a certain amount of change; even if we stopped releasing all of these gases today, the earth would still go on warming and the ozone layer would go on getting thinner for some time to come.

Ozone in the stratosphere, 15–25 kilometres above the ground, absorbs almost all the incoming ultraviolet radiation, letting through only enough for a suntan or sunburn. But even today 100,000 people die of skin cancers each year, and almost all skin cancers are caused by ultraviolet radiation (specifically the harmful bit called "UV-B").

The US Environmental Protection Agency estimates that for every one per cent depletion of the ozone layer, skin cancers would rise by two per cent. These cancers affect mainly fair-skinned – that is "white" – people. UV-B radiation can also increase the incidence of eye cataracts. Human skin may be able to develop some tolerance to such radiation; eyes cannot, and become more sensitive with each exposure. For every one per cent depletion of the ozone layer, another 100,000 people will go blind, according to a UN report.[2] Using satellite data, the United Nations estimates that the ozone layer decreased, globally, by 2.6 per cent between late 1978 and late 1985.[3]

A HUNGRIER WORLD

A warming is not just a matter of degrees. It is expected to change the global climate and local climates, wind and rain patterns, and perhaps even ocean currents. This will change farming, bringing new crops to some areas and making it impossible to grow familiar crops in others.

Agriculturally, there may be some losers and some winners. Who will be which? And how will the globe do on balance: will there be less food, more food, or will the planet just about break even?

There are no precise answers to these questions, but the first attempts to find answers have raised the alarming prospect that a warmer world might be a much hungrier world.

The UN system has created a global committee called the Intergovernmental Panel on Climate Change (IPCC) and given it the job of providing the best scientific data possible on climate change, in hopes that this will point to solutions. It was due to release its overall report in autumn 1990. Meanwhile, interim reports were leaking out from the various working groups. One of the groups, working on the impacts of change and led by Dr Martin Parry of Birmingham University, did a draft study of the impact of climate change on agriculture.[4] It noted that only 21 out of the world's 174 nations export cereals. In 1987, 77 per cent of all the cereals traded in the world came from only three nations: the United States, Canada and France. A major surprise in Parry's projections is that – despite more warming in the North – Canadian agriculture does not improve very much, if at all. Production is expected to fall in Alberta and Saskatchewan, but perhaps increase in Manitoba and Ontario. As northern Europe will become more productive and southern Europe less so, nations in the middle, such as France, also experience little overall change.

But the main shock is the United States, where "overall production is estimated to be sufficient for domestic needs, but the potential for export is reduced" under the CO_2 doubling scenario, according to Parry's group of more than 35 scientists. When the US does not export, the world suffers. In 1987/8, world stocks of wheat and coarse grain stood at 353 million tonnes. Then there was a heat wave and drought in the US Great Plains and the more southerly Corn Belt. Almost entirely due to this disaster, the world grain stocks fell to 248 million tonnes in 1988/9.

Basically the warming is expected to move the rain belts of the middle latitudes closer to the poles. This would mean drier summers in North America, Southern Europe and parts of the Soviet Union. The Great Plains of North America are expected to become even drier as snows melt earlier, exposing the ground to more sunlight, which evaporates soil moisture. As soils dry, there would be less evaporation, so fewer clouds, so less rain and more sunlight hitting the ground directly.

Production is expected to improve in China, the Soviet Union and much of Northern Europe. For instance, potato yields in Britain may well increase by 50–75 per cent. But even though Northern Europe is expected to produce 10–30 per cent more food, the facts that it imports so much of its food, and that the price of this food is expected to rise, mean that Northern Europe is left a little worse off, food-wise, after the warming.

Some areas of particularly high risk are those which already have trouble feeding themselves and which may get drier. Parry's group listed under this category much of Africa; Mexico and Central America; parts of drought-prone eastern Brazil; Arabia; and in Asia, Vietnam, Cambodia, Thailand, Bangladesh, North-east India, Burma, and parts of Indonesia.

Looking at other climatic change effects, Parry's group found that CO_2 actually increases yields of many crops, but this can be offset by warming and drying. The report did not consider increased ultraviolet radiation from ozone depletion, which can damage plants. It warned of more agricultural pests, as warmer conditions mean pests move out of their usual areas, and as longer summers cause them to mature earlier and produce more generations per season. Livestock diseases now contained in Africa – rift valley fever and African swine fever – could move into the United States. And it predicted that less rain in the United States could eventually mean less groundwater for irrigation.

Among its conclusions, the group said that "certain regions that are currently exporters of cereals may be especially prone to reduced soil water and thus reduced productive potential. Any decrease in production in these regions could markedly affect future food prices and patterns of trade. These regions include: southern Europe, southern USA, parts of South America, western Australia."

Yet it also concluded that on balance global food production could be maintained "at levels sufficient to meet world demand". This is an odd statement given the evidence in the report and Parry's own stark statement to a conference in Cairo in late 1989 that: "Increased production in higher latitudes is very unlikely to make up for reduced production potential further south".

However, even if the report is right about "sufficient levels", it may offer little comfort. World food production has been running ahead of population growth for some time, but in 1980, 730 million people in 87 developing countries – one third of the entire population – were not getting enough to eat to live active working lives. This figure continued to rise during the 1980s, and was expected to go on rising.[5]

So under global warming, there could still be enough food to feed the world – as there is today – but there could also be millions more left hungry as tighter markets meant food did not move, and as already hungry and marginal regions got drier. In the mid-1980s, one African in four was being kept alive by imported food, much of it from North America. About 70 per cent of all food aid is given or sold to governments at concessional rates; only 10 per cent goes to relieve disasters. The remaining 20 per cent is "project" aid meant for the poorest, much of this in projects to feed children. All of this aid, especially the 30 per cent, is important to children's survival. A warmer world could be one of smaller, if any, surpluses and less, if any, food aid.

OZONE AND FOOD

Crops are also expected to be hurt by ozone depletion; of 200 plant species and varieties tested under increased ultraviolet radiation, over two-thirds showed some sensitivity. They do not germinate or flower as often, and crop yields fall. But the effects of ozone depletion on crops are more pervasive.

In tropical rice paddies live species of bacteria which fix nitrogen in the crop's roots and help the plants to grow. These bacteria are extremely sensitive even to present levels of ultraviolet radiation; rises could diminish Asia's all-important rice harvests. UN scientists looking into this problem found:

> The annual nitrogen assimilation by this group of organisms alone has been assumed to amount to 35 million tons, which exceeds the 30 million tons of artificial nitrogen fertilizer produced annually. The amounts of artificial nitrogen fertil-

izer necessary to compensate for a substantial loss will stress the capabilities of less developed countries.[6]

Put simply, the poorer countries will not be able to afford the nitrogen fertilizer to replace the effects of the microbes, and may not be able to grow enough rice to feed their people.

Both increased ultraviolet radiation and sea-level rise are expected to decrease fish catches. Fish provide ten per cent of all the animal protein consumed by humans, but in some areas, especially in poor countries, this figure is much higher. Asians get 40 per cent of their animal protein from fish, not just from the sea but also from small ponds and streams and even irrigation ditches. Children are sent out with wicker traps to catch tiny fish, which are pulped and used as sauces. This is about the only animal protein millions of children across Asia ever get. A warmer world could mean fewer sources of water and thus of fish.

Most of the world's fish and shellfish are either born in or spend their early lives in coastal marshes, mangrove swamps or other wetlands. Sea-level rise could destroy many of these faster than they could be reproduced along new shorelines.

More ultraviolet radiation could also directly kill fish and their food. UV-B kills zooplankton, the animals of the plankton world near the bottom of the ocean food chain. It can also kill fish larvae and young fish, shrimp larvae, crab larvae and other food sources for larger fish. One experiment found that exposing a body of water for 15 days to UV-B levels 20 per cent higher than normal killed all anchovy larvae to a depth of 10 metres.[7]

Plant plankton (phytoplankton) may also be killed by increased radiation. They are important in the ocean food chain, and they also absorb half the CO_2 produced in the world each year. When they die, they take much of this carbon to the bottom with them. If there were fewer of them about to absorb carbon, then global warming and sea-level rise would accelerate.[8]

WARMER PEOPLE

It is hard to know exactly how any city or neighbourhood will be affected by global warming and when. Scientists think the world has already warmed about one half of a degree centigrade. US scientist James Hansen predicts that the average person in the street in Europe or North America will begin to *feel* the warming in the 1990s. It is arguable that they already do, since six of the warmest years ever recorded occurred in the 1980s.

Hansen and his team have tried to make some temperature predictions for various US cities. Today in an average year, the temperature in Washington DC rises above 38°C (100°F) only one day a year and exceeds 32°C (90°F) on 35 days. By the middle of the next century, Washington could be having 12 days over 38°C per year and 85 days over 32°C. Even with global average rises of only a few degrees, fierce heat waves could still savage cities in hotter, drier areas, as small average changes mask some huge surges.[9]

Adult human beings can of course adapt to large changes in temperature, but infants are not good at adapting. This ability rises slowly with age through adolescence, reaches a peak which continues until about 30+ years of age and then declines.[10] Children are less able to tolerate extremes of heat or cold because of their larger surface area relative to body weight. There is more area to absorb or give off heat and less bulk to maintain a healthy temperature. Babies' bodies are also less well insulated than adults'.

So both infants and old people are vulnerable to temperature extremes, hot and cold. Even in relatively mild, northern European countries, the old and the very young are at risk from dehydration during hot, dry spells. The degree of vulnerability depends on what people are used to. Death rates start to climb in New York City when the temperature rises to about 33°C (92°F), but further south in the normally hotter city of Jacksonville, Florida, death rates remain normal when the temperature climbs to 38°C (100°F).

The UN's World Meteorological Organization, whose climatologists normally issue only the most cautious and conservative statements, has said that "The effect of such temperature rises on human health in Washington and similar cities throughout the world is difficult to predict. However, there is no question that increased urban heat stress could come to claim many lives."[11]

Most of these claims will be met by lives in the Third World.

In the early 1980s, Indian environmentalist Anil Agarwal surveyed urban housing conditions in developing countries. Things have changed since then, but much of the change has been for the worse as slum and tenement populations have continued to grow. He found that in Kanpur, India, 58 per cent of all families had only one room, and three-quarters of all houses had no windows. In Tehran, 92 per cent of the squatter population lived in one-room tenements, the average living space per person being three square metres. In one Karachi squatter settlement, four per cent of the households had more than 10 people living in one room; 24 per cent had six to eight; and 22 per cent had four to six in a room.[12] Increased heat in such crowded conditions would be deadly.

As tropical cities have grown, they have already become islands of heat: heat radiates from pavements and buildings; there is less wind; and water runs quickly off pavements after rains, so there is less cooling due to evaporation. Cooling the insides of buildings by air conditioning systems pumps more heat outside, and requires more electricity, which usually means that more CO_2 is released. This heat island effect is already a problem, and may get worse.

"Cities in tropical climates can easily turn into death islands when they are too large, dense, with short distances between buildings and little green land", says WHO consultant W.H. Weihe. "Taking 35°C (95°F) as the upper critical temperature for heat-aggravated death to occur, the vulnerable age groups of . . . Calcutta, where the mean monthly maximum temperature is 36.1°C (97°F) will be particularly endangered." The vulnerable are the infants and the elderly.

OZONE, VACCINATIONS AND HEALTH

One of the more terrifying threats associated with ozone depletion is the fact that increased UV-B radiation may have an effect on the human body similar to AIDS: a degradation of the body's natural immune system.

This raises the fear that the immunization programmes against such diseases as measles, tetanus, polio and whooping cough which are now preventing two million deaths in the Third World each year could become not only less effective, but actually dangerous.

The evidence that UV-B hurts the immune system comes from experiments on animals. There is little evidence to suggest that this happens in humans, and no evidence to suggest that it does not; there is unlikely to be hard evidence unless it is seen to happen on a large scale: it is difficult to imagine human experiments that would be ethical. UNEP warned in its review of ozone depletion studies that outbreaks of infectious diseases like measles, herpes, tuberculosis and leprosy could be expected to occur among more people with greater severity as ozone levels decline.[13]

The damage process is complex, poorly understood and works in several different ways. Increased UV-B radiation impairs the workings of Langerhans cells, which are in the skin and usually surround foreign substances entering the body (antigens) and hand them over for destruction to the *helper* T-cell lymphocytes. Once exposed to UV-B radiation, Langerhans cells no longer do their job. In some experiments on people, substances meant to produce an allergic reaction no longer do so on irradiated skin, which means that the skin is not resisting attack.

UV-B radiation also increases the activity of the *suppressor* T-cell lymphocytes, which normally shut down the immune response once it has done its job. As UV-B activated suppressor lymphocytes circulate through the body, there is a general, system-wide shutdown of certain immune responses. This has been found to happen in several animal species. The system-wide disruption increases with radiation, but persists even after the radiation stops. British researchers found that UV-radiated mice injected with the virus which causes herpes had a poor immune response to the virus, a response which lasted for months. The syndrome also raises fears about diseases caused by microbes which enter the body through the skin, as does the bacillus which causes leprosy and the parasite which causes schistosomiasis.[14]

Vaccination usually involves injecting into the body a dead or weakened agent of a disease – an "antigen" to stimulate the body's defences. But UNEP warns that injecting a person through skin exposed to excessive UV-B might make that person more susceptible rather than less susceptible to the antigen. The jab could actually cause the disease it is meant to be protecting against. Given that so many children being vaccinated are already malnourished, or suffering from worms, or malaria, or all these things, this added shock might be enough to throw their systems into a downward spiral.

Research is needed to indicate which diseases will become more infectious as ozone decreases, to find out how radiation affects the immune system, and to discover whether and how the immune suppression will affect the efficiency of immunization work.

As stratospheric ozone thins, more UV-B will reach the lower atmosphere, near the ground. There it will react with pollutants already in the air and create more of the pollutants associated with smog, including ozone itself. Levels of hydrogen peroxide and acids are also likely to increase. These all lower the effectiveness of the human respiratory system, especially in children. Studies of three US cities – Nashville, Philadelphia and Los Angeles – found that a combination of ozone depletion and greenhouse warming could increase smog formation by as much as 50 per cent.[15]

There is another side to the recent increase in the pollutant low-level ozone. It may actually be taking on the role of stratospheric ozone, and protecting people from UV-B. But this must be balanced against its damage to health.[16]

WARMING AND BUGS

As the environment changes, the range of creatures in the environment which cause human diseases, their reproductive rates, and their interactions with people are all expected to change.

"Human health could be affected by even quite small changes in average mean temperature and there is a prospect of some major diseases flourishing in warmer conditions and of more resistant strains of infection emerging", concluded an expert group on climatic change gathered by the Commonwealth Secretariat.[17]

As the world gets warmer, diseases associated with the tropics – many now banished from temperate zones, if they ever existed there – may spread north and south. As temperatures rise, so may death rates from these diseases.[18] For example, the tetanus bacterium persists longer in warm, moist soils than in dry, cool climates.

Where temperature and humidity are high, viral diseases can spread more easily. Hepatitis B virus is more easily transmitted in the tropics; epidemic cerebral meningitis is associated with warm, humid air masses; and polio is more prevalent during the summer and is associated with higher relative humidity, especially in the rainy season. Cholera is another disease prevalent during the summer. Bacillary dysentery prevails in tropical countries during the rainy season. Children are the most vulnerable to all these diseases.

A warm climate may be also more favourable for the propagation of communicable diseases carried by insects and other "vectors". Changes in temperature, rainfall, humidity and storm patterns affect the reproductive rate of the vector, the number of times it bites people and the amount of time over which people are exposed to it. There are also indirect effects: climate may change agricultural systems or plant species in a given area and thus change the relationship between parasite, vector and human host.[19]

For example, malaria is carried by the *Anopheles* mosquito and is caused by the parasite *Plasmodium falciparum*, which the mosquito injects into the human bloodstream when it bites. The development rates of this parasite increase with warmer temperatures. If the number of irrigation projects is increased in a warming world, then new breeding grounds for mosquitoes will be produced.

Malaria affects over 100 nations, with over 100 million cases reported each year. In Africa alone it kills 750,000 people each year; globally death and illness rates are as high now as they were in the late 1970s.[20] Yet some areas have got worse, and others better. Over 1988 and 1989, outbreaks have been increasing in parts of Asia, Latin America and sub-Saharan Africa.[21] Mosquitoes are developing growing resistance to pesticides such as DDT and dieldrin, while the parasites are becoming more resistant – or more parasites are becoming resistant – to the anti-malaria drug chloroquine. Any development which speeds the development of the parasite will speed the development of resistance.

Warming may cause mosquitoes to move out of their usual habitats, raising the spectre of malaria moving back into southern Europe. They may also move vertically. Hippocrates advised patients suffering from malaria to move into the hills, many centuries before there was a known connection between mosquitoes and the disease. Drier, cooler highlands are naturally free of mosquitoes. But warming could move them uphill, into previously unaffected highlands in places such as Ethiopia, Kenya and the Andean foothills. Today, many Ethiopians are dying of malaria as they are moved down from the eroded highlands to the more fertile lowlands; they have no natural resistance and there is little or no provision of anti-malarial drugs. The effects would be the same if the mosquitoes moved uphill.

The snail-borne disease schistosomiasis may also increase with more irrigation and movements by large numbers of people to farm the newly irrigated land.

Fears of increases in vector diseases are not confined to the tropics; there has also been rising alarm in the United States. Lyme disease is caused by a bacterium carried by several subspecies of ticks; the ticks themselves hitch rides on several mammals, including deer and mice, and on birds. The disease, first identified in Lyme, Connecticut, as recently as 1975, causes profound fatigue, fever, chills, headache, backache and heart abnormalities. It is spreading rapidly in the United States, where the ticks mainly bite children playing bare-legged and bare-foot in woods and on lawns. It has more recently been found in Germany, Switzerland, France and Austria. The disease is at its height during the summer months of June and July.

Lester Grant of the US Environmental Protection Agency warns that "lengthening of warm weather periods and shortening of winter weather could be expected to enhance the abundance of the tick vector and its potential spread into adjoining areas". He expressed similar concern about the ticks which spread Rocky Mountain spotted fever, a disease which is often fatal unless quickly diagnosed and treated. High temperatures are needed for the transmission of this disease, much more prevalent in the south-eastern United States than the Rocky Mountains, where it first emerged.[22]

In the summer of 1988, a sudden bloom of algae off the coasts of Norway, Sweden and Denmark caused at least $4.5 million worth of damage in salmon farms along the southern Norwegian coasts alone and did uncalculated damage to natural fisheries. The poisons released by the algae killed all sea life in some places from the sea-surface to 30 metres down. Acid rain and fertilizer run-off were among the various causes put forward, but there was no proof. However, some scientists fear that such blooms will increase as oceans warm and fertilizer run-off increases.

Bigger pests may also proliferate in the oceans. One UN report warned that sharks are especially attracted to warmer waters and could turn up off beaches where they have not previously been a problem.

Diseases will move in a warmer world; so will people. Large numbers of people will flee coastal floods, more and fiercer cyclones and drought. Stjepan Keckes, who has for years run UNEP's regional seas work, has warned that global warming may cause whole populations to become "boat people". "We are not talking about the loss of islands; we are talking about the loss of nations", he said.[23]

Climate refugees will change disease patterns by taking diseases with them into areas where they may be new. "People with infections and unimmunized people are likely to appear in countries with developed health care systems, with a resulting increase in the incidence of diseases such as measles, pertussis [whooping cough] and poliomyelitis, as well as that of tuberculosis, leprosy and other chronic illnesses", warns Dr Alexander Leaf of Harvard Medical School.[24]

DISASTERS FIRST

The fact that average sea-level rise is a matter of fractions of a centimetre per year, and temperature rise of fractions of a degree, leads one to the comfortable conclusion that change will be slow and gentle. This is a dangerously false conclusion. Change will be felt in the form of disasters. We may already be feeling them.

There is absolutely no proof that the gales which struck Britain and northern Europe in October 1987 and in early 1990 either were or were not connected with global warming. The big computer climate models offer no help; they have nothing concrete to say on wind patterns around Britain in a warmer world. Some British climatologists feel that only in the early twenty-first century will we be able to look back at those storms and say, "Ah yes, those were our first warnings of what was to come".

"These storms, whatever their immediate cause, are harbingers of the kind of weather we should expect if the greenhouse effect gathers pace as predicted in the 1990s", wrote science journalist John Gribbin.[25] He points out that such storms are caused when warm tropical air moves northwards to meet cold polar air, the cold air lifting the warm, causing clouds and rain. Energy generated twists the air mass into a circular low pressure area and causes gales. As long as there is enough polar ice to keep northern air cool, and as southern areas warm, gales will become fiercer and more frequent.

The 1990 Northern European gales killed over 120 people, a far cry from the 1970 cyclone in East Bengal (now Bangladesh), which killed an estimated 240,000. Cyclones – called "hurricanes" in the Americas and "typhoons" in East Asia – are also associated with low pressure areas, although these develop over warm tropical waters; so cyclones always strike land from the sea. Cyclones are expected to increase in severity and frequency as the globe warms. About 90 per cent of deaths in such disasters are caused by drowning during "storm surges", and the rest by flying debris and falling trees and buildings.

Storm surges are an odd phenomenon, something between a tidal wave and an extremely high tide, though related to neither. The air pressure differential lifts coastal water, and winds blow it inland. It moves not at the speed of the wind, but at the speed of the storm as a whole, perhaps 15 kilometres per hour or so. Waves riding faster along the top of the surge may cause damage. The surge is stopped by high ground, but it may keep the land covered until the eye of the storm passes over – perhaps for three to five hours.[26] John Seaman of Save the Children UK has pointed out that many survivors of cyclones, certainly many of those who survived the 1970 disaster, suffer from "cyclone syndrome". This consists of severe abrasions of the chest, arms and thighs, caused as people cling frantically to trees for hours until the surge abates.[27] Anyone who has travelled recently along the southern coasts of Bangladesh can see that this option is no longer open; most of the trees have been cut down.

The increased flooding from oceans expected in a warmer world will occur as tidal extremes bring higher seas over dunes or other barriers. The Warringhe Shire Council, responsible for a community of 200,000 people and 65 per cent of the beaches of Sydney, Australia, had designed its drainage systems to take account of a severe flood expected once every 50 years. In light of global warming, it is considering rebuilding for such floods expected once every 17 years after 2030.[28]

But the poorest nations, and the poorest children in the poorest families in those nations, remain the most vulnerable as disasters increase – whether those disasters are cyclones, floods, droughts, epidemics, or famine associated with crop losses.

The effects of even the most natural of natural disasters are more determined by cultural and social factors than by the physical nature of the disaster itself. In 1972, an earthquake hit Managua, the capital of Nicaragua, affecting an area inhabited by 420,000 people, and in 1971 another less strong on the Richter scale struck the San Fernando area of California, affecting a population of seven million.

Yet the Managua quake killed 4,000–6,000 people; the California quake, 60. Some 50,000 houses were destroyed or rendered uninhabitable in Managua; 914 in San Fernando. Most victims in Managua died because poorly made houses and buildings fell on them, but other factors in the very different tolls included education on what to do in a quake, preparedness, and quick medical care.[29]

Poor nations are simply more vulnerable to disasters. Japan is disaster-prone, having suffered 43 natural catastrophes between 1960 and 1981. Peru is slightly less disaster-prone, with 31 events over that period. But the average Japanese disaster killed 63 people; in Peru, the average was 2,900.[30]

Children are most vulnerable in almost all disasters. In slow events such as droughts the children succumb to a combination of shocks such as malnutrition, diarrhoea, parasites and the rigours of moving to find food. During the three-year drought in Africa in the mid-1980s, 150 million people were affected, three-quarters of them women and children.

In the more sudden catastrophes, statistics show fatality peaks among very young children and the elderly. In the 1970 East Bengal cyclone, half of the dead were children under ten, even though they were only one-third of the population affected.[31]

There have been odd findings about children in other disasters. A study of one village following the 1976 Guatemala earthquake, which killed 23,000, found that the child most at risk of death was the second-from-youngest. Rates of death then decreased as children increased in age, as in the Bengal cyclone. The very youngest seemed to be protected by sleeping with their mothers; but in households in which mothers were killed, the youngest child usually died as well.[32]

The fact that children, especially malnourished children, are less resistant to cold than adults becomes important even in the warmer tropics as cyclones and floods increase. Danger from exposure is high after children have got wet and chilled, and dry blankets and extra clothes may be in short supply. Even in a drought, death rates of children can increase due to cold and damp, as families take shelter in flimsy huts in famine camps hit by brief but hard rains. This happened repeatedly in the Ethiopian camps during the last decade, and children were left sitting in water-filled depressions in the bottoms of huts during the cold nights of the highlands.

Aid agencies tend to rush in vaccines and begin vaccinating everyone in sight after a disaster to lower the risk of epidemics. In fact, such dangers appear to be grossly overrated, except for epidemics of malaria after cyclones when heavy rains provide new breeding grounds for mosquitoes, wash away previously sprayed pesticide, and disrupt spraying programmes.

But experts now talk of chains of disasters. For instance, a drought drives hungry people to seek food, and they crowd into famine camps in crowded, unsanitary conditions. There, the hunger problem may be solved for a time, but the camp experience becomes the new disaster as victims contract diarrhoeal diseases and dysentery, measles, whooping cough, malaria, tuberculosis, and scabies and other skin infections. Children under five suffer most and often die.[33]

WRECKING THE SYSTEMS

Sea-level rise could directly affect a surprisingly large number of people, and the complex systems that they have constructed to meet their needs.

Nearly one-third of humanity lives within 60 kilometres of a coast. However, it is not how close one lives to the sea, but how high one lives above it, which determines safety. Along the steep coastal cliffs of much of Britain and Norway, being only a few metres from the coast is enough to protect one from any probable sea-level rise. However, large parts of Britain are vulnerable, especially London, other ports, and parts of East Anglia. A Parliamentary committee taking evidence on sea-level rise asked a scientist from the British Antarctic Survey if he could see any good coming from climate change. He replied that it would be convenient to be able to sail the Survey's research ships directly to its headquarters in Cambridge.

In other countries, such as Bangladesh, the storm surges caused by cyclones already bring water 200 kilometres inland. A study of Bangladesh and the populous Nile River delta of Egypt looked at what would happen if seas rose by a maximum of 79 centimetres by the year 2050. In Bangladesh, this rise – given other factors such as natural subsidence and dammed rivers – would cause the loss of 18 per cent of the nation's land, which now supports about 15 per cent of the current population – about 16.5 million people. In Egypt, the worst case for 2050 would affect 19 per cent of the land and 16 per cent of the current population – about 8.5 million. But in both countries, the population may have increased fourfold by the year 2050, given present growth rates.[34]

In many small island nations, there is not room to move far

enough away from the sea. Pacific countries such as Kiribati, Tuvalu and Tonga might have to be abandoned. In the Maldives, one of 21 small island Commonwealth nations, very little land is more than two metres above sea-level.[35] Already large parts of the Maldives are being buffeted by storm surges. Male, the islands' capital, is only 700 metres wide and provides a home for 56,000 people. Other islands in the Pacific and Caribbean may have more high ground, but capitals – and almost all schools, hospitals, industry and people – tend to be on the coasts.

Rich nations can protect themselves against rising seas, up to a point. The United States would have to spend $160 million yearly to protect the people, houses and industry in threatened areas against a sea-level rise of one metre. Not only is this a tiny fraction of US Gross National Product (GNP), but such spending would save $2.1 billion in annual damage averted, according to one IPCC study.[36]

It is harder for poorer nations, with lower GNPs, to find or even to justify such expenditure. Coastal protection against a one-metre rise would cost Bangladesh $332 million a year – five or ten times as much as it would save, given that the protected land is not as productive as protected US land. Other nations with large threatened areas – Brazil, The Gambia, Mozambique and Vietnam – are in a similar situation.

Sea-level rise could also spread infectious disease as the sewerage and sanitation systems of coastal cities, North and South, are flooded and bacteria are released. Farmland could also be contaminated for years with sewage, or with toxic waste transported by flood waters from waste dumps.

Hazardous waste dumps are a growing problem which could become worse as seas rise. Some 375 million tonnes of hazardous wastes are being produced every year.[37] There has been a recent trend for companies in the industrialized nations to cope with these large amounts of wastes by exporting them to developing countries, few of which have the technology to deal with them. Most of this material is dumped near the sea in the Third World because it is brought in by ships and there is no way to transport it inland.

Metal drums containing such wastes could be damaged as they

come into contact with highly corrosive sea water at coastal waste sites. In the United States, there are reported to be 1,100 active hazardous waste disposal sites within floodplains, and there have already been several environmental disasters caused by flooding.[38]

Hazardous waste has proved an intractable problem in many wealthy nations. In 1981, the US Environmental Protection Agency set up a "Superfund" process to clean up the 1,077 toxic waste dumps identified as threats to local people and to groundwater supplies. By 1989, 1,047 of these still remained to be cleaned up.[39]

Such facts are freely available to US citizens. In Britain, people are not given such information. In early 1990, the *Observer* newspaper and Friends of the Earth teamed up to publicize the exact locations of more than 1,350 waste dumps in England and Wales identified by the government as threatening to contaminate water sources. They did this by publishing the results of a government survey carried out in the 1970s, but kept secret, even from the National Rivers Authority. At the time of publication, the Department of the Environment confirmed that there was no national decontamination exercise underway.[40] Many of the dumps are near the sea, but it was not clear which would be threatened by expected sea-level rise. The government may know, but it is not telling.

This is only one example of the British government's casual approach to waste full of cancer-causing agents and other poisonous chemicals. UN figures show that in Korea in 1983, only 22 per cent of all hazardous waste went into landfills; nine per cent was burned and the remaining 69 per cent was recycled. In 1985, Sweden, with its reputation for cleanliness, could not match this; it dumped 41 per cent of such waste in landfills. But in 1986, Britain was disposing of an amazing 82 per cent of its hazardous waste in landfills, burning two per cent, treating eight per cent, dumping eight per cent into the ocean and recycling none. It was not clear whether these figures for Britain covered only its own waste, or also covered the more than 55,000 tons of "classified special waste imports" brought into the UK in 1986/7.[41]

Once poisonous chemicals get into deep aquifers, there is no way to get them out. Rivers cleanse themselves relatively rapidly, once the polluting is stopped; lakes take longer, but eventually become clean. Most groundwater deposits, however, are renewed only over centuries. As people in the parts of a warmer world which has become drier seek clean water, they may find themselves pumping up poison.

CLIMATE CHANGE AND THE POOR

It should be obvious by now that, even though northern latitudes will warm faster than the tropics, it will be the poor tropical nations which will have the hardest time adjusting.

Disasters are more destructive in under-developed nations; and there will be more of them. The South is more directly dependent on agriculture than the North, and agriculture will be disrupted. There will probably be less food aid available and less food for sale. Poor nations just do not have the money to adapt to change.[42]

There are two broad types of response to climate change. Individual governments can *adapt* to change, by building sea defences, moving infrastructure threatened by the sea, developing crop varieties which can cope with the new climate, etc. Or they can *limit* the scale of change by decreasing all of the greenhouse pollutants. In practice, both sorts of measures will be necessary. The greenhouse gases already in the atmosphere commit the world to a certain amount of warming and sea-level rise, to which all will have to adapt. But if increasing populations follow a business-as-usual policy, the planet will just go on getting hotter and hotter and the seas higher and higher. Limitation will be necessary, sooner or later.

Jill Jaeger, a British geographer, looked at the implications of the adapt-or-limit choice.[43] The main difference financially is that limitation will require large sums of money to be spent immediately. Adapting, especially adaptation which is not anticipatory, which reacts to events as they occur, costs virtually nothing now, but is likely to cost a great deal in the future.

The type of spending is as different as the timing. Much of the money spent on limitation, for instance on saving energy and perhaps that spent on developing new energy sources, would be in the nature of investment which could bring eventual profits. Money needed to adapt after the event: for instance, repairing flood damage and importing food, does not yield profits. There is also the problem of surprise. We could be in for sudden shocks, such as changes in major ocean currents. Big surprises could add greatly to the costs of reactive adaptation.

Concerted efforts at limitation will cost the rich countries a great deal, as they are the countries which have the ability to research new energy systems and alternatives to CFCs. Adapting later would be cheaper, in the short to medium term, for the richer countries, but disastrous for the developing countries, where costs would be measured in deaths, damage and crop losses. Eventually, merely adapting to change would be prohibitively expensive for everyone.

Dr Jaeger's scenarios are worrying because all politicians try to put off spending money and speed up the earning of money, even when the benefits of doing otherwise are obvious. Forests are being squandered when their rational management would create much greater wealth – but over time. Deserts are allowed to spread when this spread could be stopped, at great profit – but profits earned in the future. The need to do something unpleasant, such as build a waste incinerator, is often met by the reaction called "NIMBY": "Not In My Back Yard". Some UN bureaucrats trying to ease reluctant governments into hard economic decisions on global pollution have dubbed the usual response "NIMTO": "Not In My Term of Office".

Dr Jaeger drew up her adapt/limit scenarios before the arguments over climate change really heated up, as they did during the 1989 UN General Assembly session. A remarkable battle is emerging.

THE NORTH-SOUTH CLIMATE DIVIDE

The richer industrialized nations, including Eastern Europe, contain about a quarter of the world's population, but produce about three-quarters of all the CO_2 released by fossil fuel combustion: 24 per cent from the US alone, 27 per cent from Eastern Europe, 18 per cent from Western Europe and Canada, and six per cent from Japan, Australia and New Zealand. These nations produce 98 per cent and 97 per cent, respectively, of the two most harmful CFCs: CFC-11 and CFC-12. But the South uses many of these CFCs.[44] As industry increases in the South, the South's relative contribution of fossil-fuel CO_2 will rise, although very slowly.

A number of Northern countries, led by the United States and abetted by Britain, are arguing that since global warming and ozone depletion are "global" problems, then all nations are equally obliged to co-operate to limit emissions.

But the only way Southern nations can limit industrial greenhouse gases (as opposed to those from agriculture and forest clearance) is by installing more efficient energy systems and using safer CFCs or alternatives to CFCs. Those things are made in the North – and they are for sale. Southern nations listening to this line of argument hear the following: "Look, the way we rich nations have chosen to develop is largely responsible for these global problems. Now we plan to make even more money by selling you the solutions."

Nations like China and India are not having any of this. Cheng Zheng-Kang, a Beijing University law professor who helped to draft much Chinese environmental legislation, told an international conference in early 1989: "Third World countries cannot bear the responsibility for destroying the ozone layer. If the developed countries did not use so much CFC-11 and CFC-12, we wouldn't have a headache right now." He said his government's approach to the ozone issue rested on three points: the issue of who should solve the problem; a general understanding that developing countries have other problems besides environmental ones, and an open discussion of the issues.[45]

None of these points is having much impact on the North, where leaders duck these questions, perhaps because their own position is so untenable. US President Bush, in a speech to the IPCC in early 1990 in which he was expected to take a bold stand on climate change, said: "Wherever possible, we believe that market mechanisms should be applied and that our policies must be consistent with economic growth and free-market principles in all countries". This was taken to be an obfuscated way of saying, "Let the poor, the non-polluters, pay." In her speech to the 1989 UN General Assembly on the environment, Mrs Thatcher said: "It is no good squabbling over who is responsible and who should pay".

As far as CFCs are concerned, there is little else left over which to squabble. Some scientists are fond of referring to them as a "non-problem", by which they mean a non-technical problem. There are substitutes for virtually all uses. In fact, 80 nations agreed in Helsinki in mid-1989 to phase out CFCs as soon as possible, but no later than the year 2000. They also agreed in principle to help developing nations phase out the compounds, but they declined to establish the international fund for this demanded by China, India and other Southern nations.

So what is the problem? The phasing out of CFCs by the Northern nations could be offset by their use in a few populous Southern nations. Cheng told the 1989 conference that China planned to manufacture as many CFCs as the United States by the year 2000 to satisfy the Chinese demand for fridges. Already companies in Italy and Japan, countries which have signed the 1987 Montreal Ozone Protocol limiting CFC use, are selling to China, which has not signed, obsolete refrigeration equipment which will use the harmful CFCs. The short-sighted cynicism evoked by this issue is boundless, and all this over a "non-problem". What hope does this raise of solving the much more technically difficult global warming problems?

THE NEW NORTH-SOUTH BALANCE

In a more rational world it would be obvious that global warming and ozone depletion have profoundly changed the balance of power between the industrialized and the developing world.

For the first time ever, what happens in the South poses a serious physical threat to Northern nations. If China rapidly increases its production and use of CFCs, Britons get more skin cancer and American children breathe more smog. If the developing nations follow the lead of the North, and develop wasteful and dirty energy and industry systems, then the US Midwest dries out and the seas around Britain rise. Slow rates of industrial development in the South may be made up for by rapid rates of population growth there.

According to the International Energy Agency, the world will use 50 per cent more energy in the year 2005 than today. Over that time, CO_2 emissions from Eastern European nations will increase twice as fast as those from the richer industrialized countries, and emissions from the developing countries will rise three times as fast.[46]

The initial impetus in the North to give aid to the South came from a vague sense of charity, coupled with a desire to buy up Southern allies in the cold war. Then the Brandt Commission reports of the early 1980s argued that not only did the North have a moral imperative to aid Southern development, but such development would create more markets for Northern goods and more security for Northern peoples. Neither the moral nor the mutual benefits argument has had much effect.

Global warming and ozone loss change things radically. If the industrialized nations do not make sustainable development possible in the South, then no steps they take on their own can be successful in limiting global warming and ozone depletion. This not only means steps to decrease poverty and disease and population growth, but transferring non-polluting development technologies southwards at costs the South can afford.

Eastern Europe too needs investment in non-polluting development. In fact, it is so polluted, and its emissions have such a direct effect upon Western Europe, that there may be a temptation for the West to spend all of its aid for sustainable development in Eastern Europe, leaving little for the South.

If there is any left for the South, and if the donors become fixated on the problem of climate change, it may be directed at the populous, industrializing nations such as China, India and Indonesia, with some also set aside for both the industries and forests of Brazil. Nations which cannot even threaten with their pollution or with deforestation – Bangladesh, Ethiopia, most of sub-Saharan Africa – may be left entirely out of the development game. Some of the poorest nations may become a threat to the North only as producers of millions of environmental refugees.

However, the discovery in 1989 that the annual burning off of the African grasslands produces more greenhouse pollution than the destruction of the rainforests could perhaps give Africa a new stake in the game.

The list of things nations can do on their own to adapt to global warming and sea-level rise includes things like building sea walls and changing farming patterns. But if they could co-operate in worldwide adaptation, the list would expand to include things like the development of a global food security system, while Northern nations are still able to generate surpluses; more aid to poorer nations to help them manage their agriculture and water resources better and to use appropriate technology; a greater international effort to halt desertification; and a greater commitment to curbing population growth where it hinders sustainable development.[47]

DOING SOMETHING

A "law of the atmosphere" treaty should be ready for signing by the year 1992, according to senior UN officials.

The elements of such a treaty are still to be determined. The 1988 Toronto Conference on the "Changing Atmosphere" recommended cuts in global CO_2 emissions of 20 per cent (of 1987 levels) by the year 2005, with 50 per cent cuts as a longer term target. Working papers for a conference of ministers in The Netherlands in late 1989 called for a two-phase approach, the first step being for the industrialized nations to cut greenhouse emissions by 30 per cent (from 1986 levels) by the year 2000, with a vaguer and more global second phase including a worldwide CFC phase-out, reductions in deforestation and a lot of reforestation.[48]

Negotiations towards the Montreal ozone treaty were difficult and produced a weak treaty. Yet this agreement only had to cover one class of chemicals, the CFCs, not produced to any extent in the South, and already being phased out by the major US and European producers.

An effective "Greenhouse Treaty" would cover many gases – CO_2, methane, nitrous oxide; many activities – energy production, industry, agriculture and forestry; and all nations. Various ideas being put forward include the creation of a global "carbon fund", perhaps $500 million a year, to preserve forests and meet other goals, "taxes" on carbon emissions and marketable, national "permits to pollute". Such permits have been in use in the United States for some time to control air pollution. Dirty factories pay high prices for permits; as they clean up, they sell parts of their permits to others.

If national permits were allocated strictly by population, rather than by what countries already emit, then the rich nations would have to buy a lot of permits from the poor. The United States releases about five tons of carbon per person per year; China about half a ton, Britain, 2.6 tons.[49] China would thus have spare polluting capacity to sell to the United States; but arguments will rage over where to set the "norm". As countries cleaned up, they could sell off permits. Britain and the US tend to favour this scheme, or some version of it, while others see it merely as a licence allowing the rich and dirty nations to go on polluting.

A treaty will probably emerge setting limits for individual nations and leaving it to them to determine how to meet those limits. So even with a treaty in place, there will be tremendous scope for citizens' groups to apply pressure so that governments meet environmental targets in environmentally and socially sane ways.

The British government has taken little interest in energy, other than privatizing it. Its transport policy favours cars, the most inefficient form of transport, over more efficient means. Spending on the encouragement of energy efficiency has been slashed.

The United States is a much dirtier and more wasteful nation than Britain, but it has passed a federal law setting minimum efficiency levels for electric appliances. When these come into force in 1990–92, they will essentially make illegal 70–90 per cent of appliances in US shops today; but by the year 2000, this one law will be saving electricity equal to the output of 21 large power stations.[50]

Most appliances on sale today in British shops are not even as efficient as most appliances on sale in shops in Europe. Friends of the Earth energy campaigner Simon Roberts reckons that setting energy efficiency standards based on the best off-the-shelf appliances available in 1989 would produce energy savings of 30 per cent and cut by 10 per cent UK carbon emissions from the domestic electricity sector by the year 2008.

The best thing about such moves – improving public transport and making electric appliances more efficient – is that they are not panic responses to an uncertain greenhouse effect but things that need doing even without global warming.

Better public transport cuts other pollutants besides greenhouse gases and makes big cities better, safer places in which to live and do business. More efficient appliances also cut down pollutants and save household electricity bills.

Britain has remained behind its European partners in developing a sense of urgency on the global warming issue. In March 1989, Baroness Hooper, a Department of Energy minister, told the Commons energy committee: "It will be perhaps 10 or 15 years before we begin to get a grip on the problem. In the meantime . . . we should avoid being panicked into measures which might ultimately prove unnecessary."[51] It would be sad indeed if Britain had not begun to "get a grip on the problem" until after what scientists are calling the "decade of decision" is over.

But the real irony in Baroness Hooper's statement is that it is the government, not the environmentalists, which has been advocating panic measures. Then Environment Secretary Nicholas Ridley said in late 1988: "If we want to arrest the greenhouse effect, we should concentrate on a massive increase in nuclear generating capacity". In her environmental speech to the UN General Assembly in 1989, Mrs Thatcher listed three things that nations could be doing individually and in groups to improve the environment in general: river clean-ups, agricultural improvements and "the use of nuclear power which, despite the attitude of so-called Greens, is the most environmentally safe form of energy".

She did not mention energy efficiency measures, which the World Commission on Environment and Development describes as "the cutting edge of national energy policies for sustainable development".[52] She did not mention new and renewable energy sources – solar, wave, wind, etc. – which the Commission said "should form the foundation of the global energy structure during the 21st Century".

Thatcher boosted nuclear power, which indeed does not release CO_2; but energy efficiency measures have been estimated to be four or five times more cost effective in reducing Britain's CO_2 emissions than building new nuclear plants.[53]

The World Commission included members from the United States, the Soviet Union, Japan and other nations with nuclear power plants, and most of them are personally committed to nuclear energy. But after months of fierce debate on the issue, which occasionally involved screaming across the table, the commission concluded, *unanimously*, that, "The generation of nuclear power is only justifiable if there are solid solutions to the presently unsolved problems to which it gives rise".[54]

Those problems – no proven disposal system for toxic nuclear waste, high costs and uncertainties over de-commissioning plants, and the potential for low-probability but enormously destructive accidents – also forced the government to remove nuclear stations from energy privatization plans once the investors had cast a cold eye on the economics of nuclear-generated electricity.

It is not the time for panic measures, but it is time to begin to do the obvious. Stephen Schneider of the US National Center for Atmospheric Research notes that so much which should be done can produce "tie-in" benefits, even if the climate does not change as forecast. He lists among these: energy efficiency, developing alternative energy sources, revising water laws, searching for drought-resistant crop strains, and negotiating international agreements on trade in food and other climate-sensitive goods.[55]

But such things still cost money and require changes in the ways in which present governments think and operate. The money will never be spent, and the changes will never occur, unless people let their governments know that they think it is time to invest in the future.

7
CHILDREN AND THE ENVIRONMENT: UNDERVALUED RESOURCES

> The neglect of children is part of a broader pattern of neglect that includes the reckless exploitation of natural resources, the pollution of air and water, and the willingness to risk "limited" nuclear wars as an instrument of national policy.
>
> *Christopher Lasch*
>
> A week is a long time in politics.
>
> *Sir Harold Wilson*

"The United States is nineteenth in the world in infant mortality rates, even though we have the highest gross national product in the world. We have the highest low birthweight rate in the industrialized world. America is lagging behind some Third World countries in immunization rates. When people from the developing world hear me talk, they are absolutely shocked."

The speaker is Marion Wright Edelman, who is used to shocking people, mostly US people, about the way her country treats its children. A black lawyer, she began her career in the civil rights movement, but in 1973 established the Children's Defense Fund in Washington, DC; it is a private research, education and lobbying group which provides a voice for US children, especially poor and minority children. She describes its mission more simply as "To make this country see kids, hear kids, and feel kids, and invest in them before they get sick, drop out of school, get pregnant or into trouble".

"I mean, politicians always kiss kids in political campaigns. Everybody is for kids except when it comes to budget time and when it comes to making hard political choices.

"We saw in the 1980s the largest military build-up in peacetime history. Our nation invested almost $1.9 trillion in new weapons of death while cutting more than $40 billion from programmes for poor children. In fact, we've had a transfer in money from the neediest children over to the military."

Aside from the moral issues, Wright Edelman points out that the cuts do not even make hard economic sense:

"It's a crazy budget policy. Every dollar we invest in immunization saves $10 at the other end. Congress is debating whether or not to invest about $2 billion in setting up a long overdue, quality, affordable child-care system, when every dollar you invest in adequate child-care saves $4.75. Traditionally we have thought we had some children who were expendable: poor children, black, brown. A decent society should not hold back any child, but with today's shrinking numbers of children and our needs to produce a work force to compete with a united economic community in Europe and against the growing Japanese and German strength, we absolutely, economically, cannot afford to waste a single child.

"We have to turn this around by organizing and complaining and fighting very hard. We simply have to make the cause of children powerful. I think that there is a great bubbling up of concern about the state of children all over the world."

Asked about the environment, Wright Edelman said she had just joined the advisory board for "Earth Day 1990", and that such concerns fitted neatly with her campaigning for children:

"We've got to come together to talk about preserving the American future, which means its children, its families, its human resources, but also its natural resources. My son has just finished a paper on toxic waste; we are burying things to leave a time bomb for future generations of kids. I look forward to the point where we can ask every element of American society to think about the impact of their actions on children in an environmental context and a human context."

POLITICIANS, CHILDREN AND THE ENVIRONMENT

Why do we and our governments routinely and systematically undervalue both our children and our environment, which are together the basis of all future progress? The reasons are many, varied, and have to do as much with psychology as with economic and political institutions.

The root cause lies in the very nature of nations and their governments. All nations, from the most democratic to those run by corrupt military dictators, are ruled by elites, those who either began at, or have risen to, the top of the social order. Such people do not suffer directly from environmental destruction; they live in capital cities; their retreats in the countryside are both well guarded and well landscaped. National leaders never suffer directly from dirty air or unhealthy water; nor do their friends and associates. Their children do not suffer from cutbacks in education, health care, or food supplement programmes; nor do those of their friends and associates. This is as true in Europe and North America as it is in Africa, Asia and Latin America.

Governments do not, cannot and should not assume all responsibility for care of children and of the environment. Much of this is properly left to ordinary citizens. But the stark truth is that government spending on children is of most benefit to the poorest – children of parents who cannot afford to invest much in feeding them, educating them, or keeping them healthy. It is of least personal benefit to the leaders and the well-off, the needs of whose children are usually taken care of privately, including private schools not funded by the state.

Thus while there may be a large proportion of the population demanding more government investment in children, the powerful are not among them. Imagine how quickly the quality of state education would improve in any nation if a prerequisite for high public office was that the office-holder's offspring had to attend state schools.

By the same token, state spending on the environment is of most benefit to the poorest and the children of the poorest. The elite are always the most thoroughly insulated from environmental squalor and degradation. They do not live in shacks next to garbage dumps or dangerous factories; they do not depend directly for their livelihoods on natural resources; they do not even use public parks and forests.

In his best-selling book, *The End of Nature*, Bill McKibben bemoans the fact that "*we have ended that thing that has, at least in modern times, defined nature for us – its separation from human society*" [italics by McKibben].[1]

This is the view of a middle-class American writer who enjoys a ramble in the forests around his rural home after a hard day at the word processor. His view of nature is meaningless to the majority of the people with whom he shares the planet. Try telling farmers in southern Bangladesh at the height of a typhoon that their "society" is separate from nature; indeed, try that argument on Yorkshire sheep farmers when lambing time coincides with a late blizzard, or on Midwestern US wheat farmers in the middle of a drought.

But McKibben's view is shared by most rulers and legislators around the world, who see nature and natural resources as separate from their own sphere of activities. This can lead to horribly short-sighted policies, especially in nations where a high proportion of the national income depends on agriculture, and thus on a healthy rural environment.

Three-quarters of the population of sub-Saharan Africa live in the countryside. But in the late 1970s and early 1980s, Africa was the only region in the world decreasing government spending on agriculture.[2] This policy doubtless did not meet the approval of the majority of the population, but they were not consulted. It has proved disastrous for the children of Africa, because their families rely for so much of their daily survival directly on the environment, on nature. Their food comes from local fields, their water from streams and wells, their fuel and building materials from trees.

The most graphic environmental destruction among industrialized nations has occurred in Eastern Europe and the Soviet Union. "Central planning" has been blamed for production quotas always taking precedence over concern for poisoned air, soil, water and people. Indeed, the planners at the centre were far removed from the immediate effects of the pollution.

But the problem was not lack of information about the effects of their policies. In the early 1980s, the Polish trade union movement Solidarity made available to journalists many of the reports on heavy metal pollution, water pollution, acid rain, and the resulting still births, birth defects and ill health among the nation's children. These had been written by doctors, public health officials and scientists and quietly filed away in the offices of bureaucrats around the country.[3] The sudden availability of information led to a public outcry, which was stifled by the military take-over of the Polish government in late 1981. Today, that outcry has resumed, and can be heard across Eastern Europe and the Soviet Union.

Central planning was not in itself at fault. Meeting the needs of future generations will doubtless require a high level of central planning in all nations. But it will also require more co-operation between the decision-makers and their broad constituencies. The problem in Eastern Europe was that the central planners were answerable only to the few above them, not to the masses below them, who were not consulted.

Democracy makes a difference in meeting the needs of children and the needs of all people for a healthy environment. The only nations which made any real progress in cleaning up their environments during the first Green wave of the late 1960s and early 1970s were the industrialized democracies of Western Europe and North America. They had the money for the clean-up, and a critical mass of people who had the time to focus their attention on the environment and did not need to worry too much about living wages, food, clean water and adequate sanitation.

But they also had a free press, the freedom to form pressure groups and the ability to "vote the rascals out", if their elected officials did not react to environmental concerns. So when there is enough public concern, even leaders not particularly concerned about the environment may be pushed into action. The powerful US Environmental Protection Agency was established under Richard Nixon. The British Department of the Environment was established under Edward Heath. Neither man was famous for his environmental concern.

Yet even when democracy accurately reflects the concerns of the people, it is not a guarantee of wise investments in the future. Most people also have short-term hopes and fears, often focused on narrow, local concerns. Democracies can look to the needs of the future only when the vision of voters broadens to include planetary concerns and the needs of generations yet unborn.

POLITICAL PRIORITIES

The priorities of governments can best be seen by where they spend money. A disproportionate amount goes to the military, especially in the Third World. The reasons are perhaps partly historical; one of the primary functions of the modern nation-state has been to organize a common defence against others. Thus military spending has traditionally been a sacred cow, with the public rarely demanding rational justification for a new weapon.

Spending on health and education, a more recent concern of national governments, is more carefully scrutinized. Economists and experts demand to know if investments in these fields are getting the most returns. Spending on the environment, a very recent concern of governments, raises fiercer debates and fiercer demands for scientific and economic justification. Politicians demand that scientific uncertainty be eliminated before even the most obvious environmental actions are taken, yet massive scientific uncertainty surrounds the unproven technologies of the US Strategic Defense Initiative ("Star Wars") anti-missile programme, for which tax-payers are paying billions of dollars each year.

Thus in the years between 1960 and 1987, the world's governments spent $17,000 billion dollars (1986 US dollars) on their military, 13 per cent more than the $15,000 billion spent on all education budgets and 70 per cent more than the $10,000 billion spent on health care.

Opposite A girl washing in Sri Lanka. Since it is hard to put a value on children or environmental resources, governments find it difficult to invest in either. *Mark Edwards/Still Pictures*

Global populations rose spectacularly in that period, from three billion to over five billion. Yet despite the extra millions of children needing health care and education, military spending stayed and remains ahead of spending on health and education. Throughout the 1980s, 29 nations spent more on their military than on education and health combined. These included most Asian nations, China and the USSR, but also some of the poorest African nations: Chad, the Sudan, Ethiopia, Angola and Uganda.

An additional 61 nations, including the United States, spent more on military than health care. The Northern industrialized countries were spending half of all government research and development funds on the military, that total exceeding all research spending on socio-economic needs, including health care, nutrition, education, new sources of energy and the environment.[4]

It is much more difficult to quantify and compare spending on the environment and on the managing of environmental resources, as this is not as centralized as military spending. (Much of the spending on health and education is also done by private citizens.) For example, in 1986 the US government spent about $273 billion to defend against foreign military threats, few of these defined or identified. In that same year, only $78 billion was spent to limit environmental pollution, a direct threat to US citizens. Of that figure, $60 billion came from private funding by corporations and others.[5]

In early 1990 the administration of President George Bush presented its first budget, for 1991, to the US Congress; it is revealing to examine how this president, who wants to be known as an environmentalist and an educator, intended to spend his people's money.

He requested that spending on defence research and development rise by four per cent while civilian R&D climb by 12 per cent. These changes would mean that the government wanted to spend $41.4 billion on military R&D and $26.7 billion on all other R&D. The Star Wars budget has increased to $5.4 billion. The Environmental Protection Agency was to get a total of $449 million for all of its research. The Administration wanted to spend $1 billion researching global change; it also wanted another

Trident submarine, at a cost of $1.4 billion. Overall, the Bush administration planned to spend 21 per cent of federal funds on defence, down from 23.7 per cent in 1990. But one submarine is still seen as buying the nation more "security" than a deeper understanding of rapidly changing global ecosystems.

CHILDREN AND TREES

Children and environmental resources share several attributes which make it hard for political leaders to justify investing in them. Spending on either must tie up today's money in hope of returns which will not appear until relatively far in the future. It is hard for a democratic government, which must justify its existence every few years in an election, to make investments which may not bear returns for 20 years or more.

Most Third World governments, including those that do not face elections, have a more precarious day-to-day economic existence and are much more concerned with satisfying elites such as the military and the small business class. They have an even harder time justifying a project such as planting trees for soil and water conservation. Those trees will not provide benefits for at least a decade.

If, as Sir Harold Wilson said when he was prime minister, "a week is a long time in politics", then the concerns of future generations are so far over the political horizon as to be completely invisible. The reality of governments' focus on the short term, on yearly savings, flies in the face of the expressions of concern of their leaders. In 1986, Margaret Thatcher said that if only Britain could cut its £35 billion annual fuel bill by 20 per cent, this would release £7 billion for spending elsewhere, creating more purchasing power and more jobs. In 1988, she said that energy efficiency was "crucial" in combating global warming. Yet in the meantime, the budget of the government's Energy Efficiency Office has been cut from £26 million in 1986–87 to £15 million in 1989–90, with official predictions of a cut to £10 million in 1991–92.[6] Does one believe the priorities spelled out in speeches or those revealed by the budget? None of this is meant to imply that the Conservative government has a monopoly on short-sightedness; but it has been around so long that it offers the most striking recent examples.

The question "what is it worth to a nation to produce a generation of children which is healthy and well-educated?" tends to receive answers such as "a lot". But exactly how much? How does a government decide how much to invest in children, as opposed to alternative investments, when it is so hard to quantify the benefits?

The World Bank has tried to come up with some numbers, at least for the Third World. Their studies show that investment in education yields consistently higher returns than most other types of investment. For example, four years of primary education is associated with an average ten per cent increase in the productivity of the educated farmers.[7] But these returns are realized in the future, and benefit others besides the investors.

It is harder to demonstrate the exact economic returns of primary education in the North, where such education is free and compulsory. In the South, where many children still do not attend primary school, it is easier to prove the increased productivity of those who do attend.

During the 1960s and 1970s, economists churned out countless studies trying to show the links between education and national economic growth. These have dried up recently. Professor Mark Blaug of the University of London's Institute of Education in London says that this has happened:

> because the economists waxed eloquent but could not deliver hard facts. Relationships between education and the economy are too loose. The economy depends on so many other things. Many argue passionately that the relatively low numbers of British children educated beyond age 16 mean that Britain cannot compete effectively with the United States, Germany and Japan, where more children get higher education; but no one can prove it.

Thus arguments for improved education must remain, perhaps appropriately, based on moral judgements and common sense. Yet this is dangerous in nations where the ideology of governments demands economic and "profit" justifications for all spending.

It is equally difficult to quantify returns on investment in environmental management. As with spending on children, it is much easier to point to the misery and hardship produced by lack of such investments. How can one put an economic value on a predictable climate for agriculture or on the protection the ozone layer offers?

Where attempts to weigh gains and losses in environmental clean-up have been made, the findings have been surprisingly positive. In the early 1970s, both Northern governments and their industries worried that money spent obeying environmental regulations would depress investment, growth, jobs, competitiveness and trade. But a 1984 international survey found that environmental measures over two decades had actually improved economic growth and employment in many countries. The benefits to health and property and in damage to ecosystems avoided had been significant. More important, the value of the benefits usually exceeded the costs.[8]

One basic problem with attaching economic values to environmental services is that so many of them have been seen to be free: from firewood gathered in the African bush to the use of the planet's atmosphere as a receptacle of industrial waste.

"The elementary theory of supply and demand tells us that if something is provided at a zero price, more of it will be demanded than if there is a . . . price", according to British economist David Pearce. "For example, by treating the ozone layer as a resource with a zero price there never was any incentive to protect it. Its value to human populations and to the global environment in general did not show up anywhere in a balance sheet of profit and loss, or of costs and benefits."[9]

Children are "free" in a slightly different sense, in that the act which creates them generally requires little initial investment; costs arise in their upkeep and maintenance. Economics, at its mathematical and inhumane extremes, can provide some surprising conclusions about investment in children. For example, it can be argued that in an expanding economy, little should be invested in children's welfare because as adults they will enjoy a higher standard of living than their parents.

Cost-benefit studies of local and national environmental

spending are recent, but are not before time. For example, the benefits of US air and water pollution controls were found to have avoided damage totalling $27 billion in 1978, ranging from a saving of $17 billion avoided in damage to human health due to air pollution, to $100 million saved because control of water pollution improved waterfowl hunting.[10]

Looked at carefully, even environmental resources such as bogs and swamps can be shown to be worth billions; but they are rarely looked at carefully by economists. In the early 1980s, the US Army Corps of Engineers wanted to build reservoirs and walls along the Charles River and surrounding marshes near Boston to prevent flood damage. But they found that if they simply protected the marshes and other wetlands along the river, these would absorb floodwater. In fact, the value of those wetlands was reckoned at $1.2 million per year – the difference between annual flood losses given present wetland areas and projected flood damage by 1990 if 30 per cent of the water storage ability were lost. Such studies are rare, and those that exist have not prevented environmentally disastrous, and expensive, flood control structures across the United States.[11]

Economist Dennis Anderson took a long look at projects which encouraged farmers to plant trees to halt wind and water erosion and to protect crops. He produced a persuasive array of graphs and balance sheets to show that such work in Africa, when done well, had produced large profits for farmers. But he concluded that

> the recommended policies have not been applied on a significant scale in Africa because public recognition of the problems and a commitment to addressing them have been lacking. Part of the emerging tragedy is that the resources required would be small in relation to the prospective economic gains from wood production and from rising (as opposed to declining) soil fertility.[12]

So it is hard to invest in the future even when profits within a relatively few years can be virtually guaranteed, when they will go to the local people doing the work, and where the need is obvious and immense. It is infinitely harder with a challenge such as global warming, where uncertainty is high, the problem global, the benefits ambiguous and going mainly to people yet to be born.

WHO SPEAKS FOR THE FUTURE?

While the well-being of today's children and today's environment suffers from poor representation in the corridors of power, the well-being of future generations and their environment suffers from having almost none at all.

The governments of even the best-run democracies tend to base decisions on pressure from various factions: the business community, big industries, motorists, farmers, unions, and other powerful lobbying groups. None of these represent, with any force, our grandchildren. Neither does any economic system respond to their needs.

Brazil's best-known environmentalist, José Lutzenberger, maintains that the problem with "market solutions" to this challenge is that certain key players cannot participate in the market. Those left out include future generations, as well as all non-human species. The chemical and agricultural industries of the late twenty-first century might be willing to pay a fortune for the genetic diversity we squander today in destroying the Amazonian rainforest. But they have no way of making this generation an offer.

When concern for children today expands into concern for future generations, economic and legal issues become complex and profound and wander into areas of ethics and philosophy.

The basic issue can be summed up under the relatively new notion of "inter-generational equity". How can we guarantee equity between generations, so that those coming after have enough resources to achieve qualities of life similar to those which have gone before? It is a difficult issue to comprehend. Not only do the amounts and types of resources change, but so does the technology with which they are used, and so do ideas of quality of life. Does this generation of five billion enjoy a better *average* quality of life, and better access to resources, than the one of less than two billion people at the beginning of the century, given that millions of people today do not have enough food or safe water?

There is growing inequity among nations; the rich nations are getting richer much faster than the majority of the poor nations, despite endless debating between governments and in bodies such as the United Nations. Men and women can meet and argue over sexual equality. Races can debate and struggle for racial equality. Yet all of these equity problems remain far from solved.

Generations cannot negotiate, one with another. They cannot meet. While actions in the present can affect the future, it is hard to see how future people can help or hinder the present. Joseph Addison noticed this long ago, when he complained: "We are always doing something for posterity, but I would fain see posterity do something for us".

LAW AND ECONOMICS

Edith Brown Weiss, a US professor of international and environmental law, has been trying to develop principles upon which international laws of inter-generational equity might be built.

One major problem is that such laws have to be universal, because the major threats to future people are international and because national borders and ideas of sovereignty are changing too fast for one nation to protect its own distant progeny. The task of establishing inter-generational equity "must be carried out by humanity, by humans as a species. At this level, obligations and rights cannot remain national or even inter-national; they have to be planetary", noted United Nations University Programme Director Edward Ploman.[13]

Weiss maintains that while we may not have *done* very much for future generations, there has been a great deal written into the documents of the United Nations and other global agencies about the needs of the future. The charter which established the United Nations in 1945 begins: "We the peoples of the United Nations, determined to save succeeding generations from the scourge of war . . .". The preamble to the Stockholm Declaration on the Human Environment also shares a concern for posterity: "To defend and improve the environment for present and future generations has become an imperative goal of mankind – a goal to be pursued together with, and in harmony with, the established and fundamental goals of peace and of world-wide economic and social development."[14]

UN Secretary-General Javier Pérez de Cuéllar brought together concerns for both children and for future generations in discussing the UN Convention on the Rights of the Child, agreed by the UN General Assembly in late 1989:

> The way a society treats its children reflects not only its qualities of compassion and protective caring but also its sense of justice, its commitment to the future and its urge to enhance the human condition for coming generations. This is as indisputably true of the community of nations as it is of nations individually.[15]

Weiss proposes three basic principles of inter-generational equity:

- each generation is required to conserve the diversity of the natural and cultural resource base, so that it does not unduly restrict the options of future generations, and each generation is entitled to diversity comparable to past generations;
- each generation is required to maintain the quality of the planet so that it is passed on in no worse condition than it was received, and should be entitled to a quality of the planet comparable to that enjoyed by past generations;
- each generation should provide its members with equitable rights of access to the legacy from past generations.[16]

The third principle makes it clear that justice between generations is even harder than it looks, because it must be based on justice and equity *within* any one generation. When billions are kept poor and forced to over-use their soils and water to grow enough to survive – as in our own generation – then it becomes impossible to keep environmental resources intact to pass on to the future.

Also, justice between generations involves both duties and rights, the right of any generation to receive as much from the previous generation as that mass of people received from its predecessors. Present laws, and certainly present patterns of environmental destruction, guarantee neither that those rights can be assured nor those duties fulfilled. In fact, present population growth rates guarantee that there will be fewer environmental resources, per person, in the next generation. So limiting population growth becomes one of the first duties for this generation to assure the rights of those to come. The World Commission looked for a solution outside present laws and institutions, suggesting that a beginning might be "the designation of a national council or public representative or 'ombudsman' to represent the interests and the rights of present and future generations and act as an environmental watchdog, alerting governments and citizens to any emerging threats".[17] The UN Convention on the Rights of the Child has taken a similar approach in establishing a Committee on the Rights of the Child to which nations signing the Convention must report on their progress in implementing the pact.

Edith Brown Weiss has called for the establishment of a "Declaration of the Planetary Rights and Obligations to Future Generations" to deal with global warming and other issues affecting the future.

It is discouraging that in the late twentieth century we need to be thinking about setting up special bodies outside normal channels to protect the two pillars of all future progress: children and the environment. That governments have trouble dealing with the legal rights of children is shown by the fact that it took their delegates ten years of tough negotiations before they could agree on a convention guaranteeing such rights as health, education, freedom of expression, and the right to be given a name at birth. The best that can be said of this exercise is that the delegates took it seriously.

The challenges for economics are summed up in Lutzenberger's remarks that there are no "market mechanisms" to cover the needs of future generations. When classic economics began two centuries ago, it focused on the three areas of land, labour and capital, and worried about what happened when too many people began to farm too little land. But colonialism meant that many of the raw materials needed for conversion into finished products came in from foreign lands. Late Victorian economists thus began to ignore natural resources.[18]

This oversight continued up until the present day. Economists, where they have dealt with environmental damage at all, have usually referred to such an effect as an "externality", something outside of the focus of the given exercise. Now a growing number of economists are trying to establish ways of "internalizing" those externalities, of working the costs of environmental damage and the benefits of improving the environment into all economic planning.

The World Commission suggested that governments establish an annual audit of environmental resources and environmental quality to complement the annual budgets and economic development plans. Nations such as France, Norway and Japan have all made stabs at this, with varying degrees of success. Yet such exercises are valuable even if their only immediate effect is to show that economics and the environment cannot be separated. Professor David Pearce, adviser to the British Department of the Environment, argues that "environmental policy matters not just for the quality of life in general, not just because natural environments have values 'in themselves', but because environments and economies are not distinct. Treating them as if they were is the surest recipe for unsustainable development."[19]

But what of children? How do governments account for their well-being? Governments go to a great deal of trouble regularly to check the health of the economy, to find out what and how much is produced, how much is saved, how much is imported and exported, how many new houses are built. It is a very costly business, but it is done because it is seen to be important. Most governments know more about the condition of the economy than the condition of children: how many are in school, how well they are learning, what diseases afflict them, how well they are eating. These "indicators" of national success are rarely seen to be as important as economic indicators.

The legal and economic tools are becoming available to offer real protection for the environment, for children and for future generations. But they will not be used until the ordinary people represented by governments demand that they are used. This can happen only if more people have more knowledge about how children and the environment are being misused.

"Failure to protect the physical, mental and emotional development of children is the principal means by which humanity's difficulties are compounded and its problems perpetuated", argues UNICEF. "The essence of civilization is the protection of the vulnerable and of the future: children, like the environment, are both the vulnerable and the future".[20]

8
REMEMBERING THE FUTURE

> In our condescending haste to teach the Eastern Europeans lessons, we may forget that democracy is more than a procedure for choosing one's rulers. It is a way of life.
>
> *Michael Ignatieff*
>
> Our children will not know it's a different country. All we can hope to leave them now is money.
>
> *Philip Larkin*

"Neither our well-off nor our poor children have any childhood", complained Dr Anandalakshmy. "The poor children are working as labourers to earn money or working in the home to help the family survive. The well-off children are in school, working for grades. There is no joy, no spontaneity. There are no longer even any games, only 'athletic events' where children compete for gold or silver."

Dr Anandalakshmy, once a well-off child herself who grew up to take a PhD in psychology in the United States, knows the problems of poverty. She is principal of Delhi's Lady Irwin College and chairperson of the "mobile creche" movement in India, which cares for the children of the women who build India's buildings. Most of these women have no permanent homes, but migrate from place to place, often city to city, looking for work. There is no place to leave the babies, except on the construction sites. So the creches, too, are mobile.

"I'm sure I am romanticizing my own childhood. And I suppose you'll say that none of this is connected to environmental destruction, but I think it is. This competition, this industrialization – it's just every man for himself, to do the best, to get ahead.

This has taken hold of people everywhere. It takes out the humanizing effect, the concern for what humans really need."

"Children are invisible to governments; we have no mechanisms in our planning for children", said Dr Anandalakshmy, who heads various government committees on children's welfare. "I was asked once what could be done about this, and I said, 'Let's lower the voting age to three. Let the children vote for what they want.' I was joking, of course, but I don't see any other way to get governments to notice children."

"The children, buffeted about, they have no feeling of control over their lives. This is the kind of tragedy which grows inside. And so when they grow up, how is it possible to expect them then to have any control over their own lives or their societies?"

"But when we talk of these issues – children and poverty and the environment and everything – we all talk so globally that anyone who wants to do anything says, 'I can't handle this; it's too big.' So I have always encouraged those who come to me or come to our meetings to handle the size of the problem they can. But start *something*, rather than worry about the whole problem."

WHERE ARE WE GOING?

It is easy to pull together a few current trends – many of the trends discussed in this book – and paint a picture of a twenty-first century hell on earth.

The present population is already consuming resources in ways and amounts that cannot go on; it is polluting the atmosphere in ways that will disturb societies, ecosystems and people everywhere. Yet the present five billion is expected to double, perhaps almost triple, in the coming century. The poor are becoming poorer in many parts of the world, and more numerous almost everywhere.

The poor nations are shipping more money than their economies can stand and more commodities than their ecosystems can stand to the richest nations. No serious plan exists to cope with the debt crisis, or with stagnant or falling commodity prices.

Political leaders remain locked in short-sighted, narrow

concerns, focusing on "sovereignty", despite growing evidence that it no longer exists. The borders of sovereign states are porous to pollution, to information, to economic shocks caused by events in other nations, to modern weapons, to refugees and to diseases such as AIDS.

But these are political, not technical, problems. The technology to solve the key problems exists. Enough food can be grown. The medical technology exists, often cheaply, to keep people alive through the vulnerable childhood years. Pollution can be controlled – though there remain valid arguments over how to use that technology, and when. The information needed to cope with such problems can be gathered, sorted, stored and moved where it is necessary.

There is also growing evidence that more people are ready to care for the environment and for children. Concern for the environment is rampant. A 1988 Gallup poll in Britain found people as much concerned over "poisoning of the environment" as over global war or nuclear destruction. A 1989 Gallup poll in the United States found that for the first time in more than a decade a significant proportion of the population considered environmental problems the most important problems facing the nation. Three-quarters of Americans now think of themselves as environmentalists. The poorer were more concerned than the richer over air pollution and toxic and nuclear wastes, "perhaps because of the greater likelihood they may live in neighbourhoods where these problems could directly affect them".[1]

Across Europe, people of one nation after another list the environment as their major preoccupation. In Denmark, there are more members of environmental groups than there are Danes, because so many people belong to more than one group.

Overleaf Children line up for a charity meal in Calcutta. *Inset* British children go shopping. The quarter of the world's population which lives in the North consumes three-quarters of the planet's resources, leaving a quarter of the resources for the three-quarters of the population which lives in the South. *Mark Edwards/Still Pictures (Calcutta photo) Lawrence Moore/Central Independent Television PLC (British photo)*

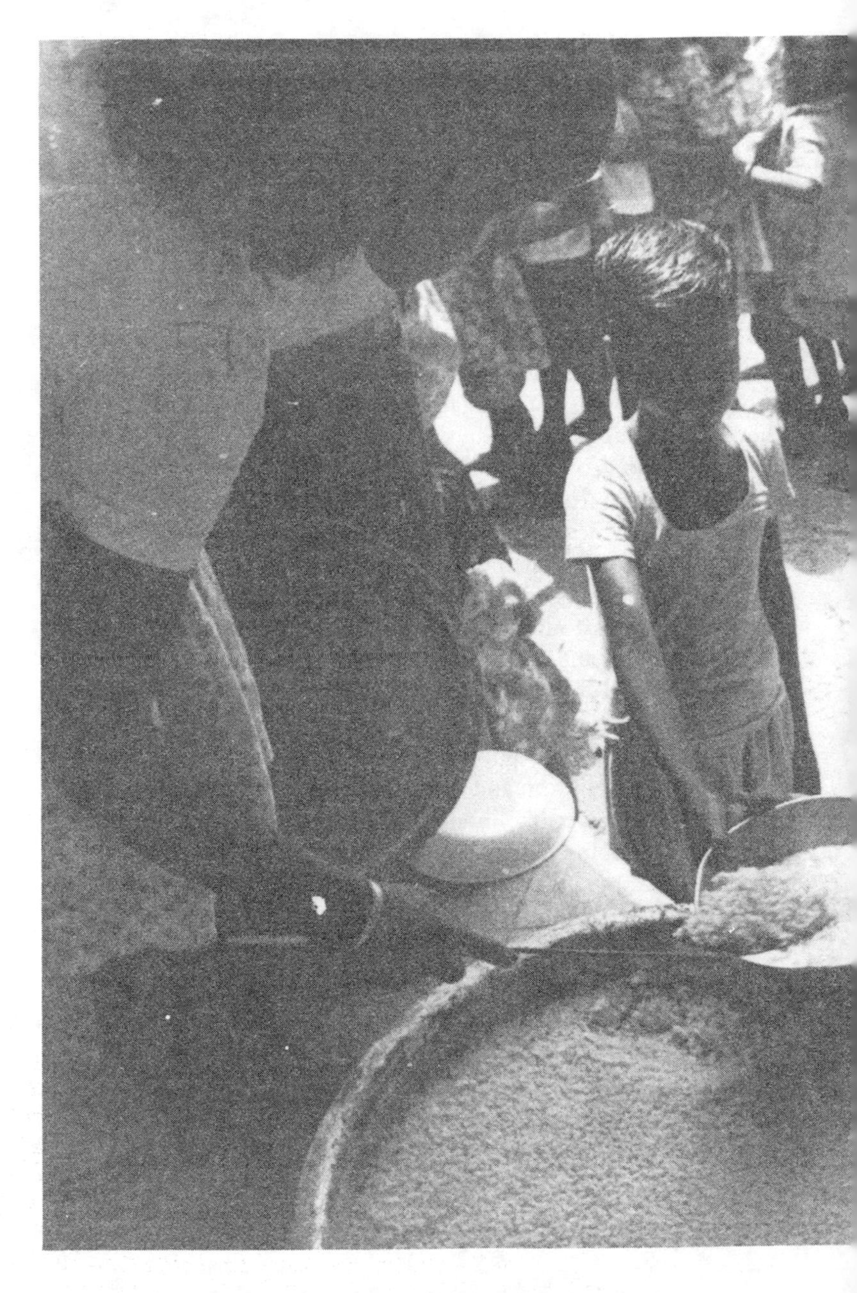

People in poor countries are also worried. A 1988 UN-sponsored Harris poll in 14 countries, nine in the South, sought the opinions of both ordinary people and leaders. It found that large majorities – between 75 and 100 per cent in all countries – agreed on the need for stronger actions by the governments, by international organizations, and for stronger laws to contain industrial pollution. Most people and most leaders were found to be pessimistic about both the five-year and the 50-year outlook for their environments.[2]

All of this represents an outpouring of popular concern, with which political leaders have struggled to keep up. "What is happening is more fellowship than leadership", said Stanley Clinton Davis, former European minister for the environment.[3]

The challenge is to get this concern out of the opinion polls and into the governments which are responsible to the concerns of their electorate.

WITHER DEMOCRACY?

The larger environmental issues, particularly global warming and ozone depletion, pose a challenge both for human survival and for the survival of democracy.

The issues are complex; yet unless ordinary voters can comprehend them, there is a danger that democracy will give way to technocracy, in which scientists and engineers decide how societies must be restructured to cope with these challenges – rather than the people who make up those societies.

Already democracy is struggling in nations where it has existed for longest and which gave it its present forms. Commentators both inside and outside Eastern Europe have noted that the recent popular revolutions there serve to emphasize how apathetic the citizens of Western Europe and North America have become in guiding the destinies of their children, their environment and their nations. In the United States, less than half the electorate bothers to vote in national elections.

Britain's political and philosophical pundit Michael Ignatieff blames the decline in political parties and in trade unions for much of the decline in democracy: 'Across Western Europe, parties die between elections and return to life largely to finance media campaigns directed at the ever more disaffiliated floating voter", he wrote. In Britain, Germany and France, union leaders have concentrated more on delivering block votes to parties than on empowering their members to participate in the democratic process.[4]

The World Commission on Environment and Development listed several preconditions for sustainable development. The first is "a political system that secures effective citizen participation in decision making". This does not guarantee a safe future, but a safe future cannot be struggled towards without it. Relatively few nations have anything like such a system, and in those farthest from such a system both environmental degradation and infant mortality and malnutrition remain rampant: South Africa, most other African nations, Nepal, Bangladesh, El Salvador, Haiti. The list goes on. Such nations will never develop sustainably without revolutions of one sort or another.

But what will become of the most powerful Western nations, if their citizens participate less and less in decision making, just when those decisions are most difficult and most important?

GETTING THE FACTS

One prerequisite for participation is information. Most Western nations have freedom of information laws. Britain does not.

When former *Sunday Times* editor Harold Evans, now working in the United States, visited Britain recently, he chided the government for thinking that its information was the property of the government, not the people. He noted that the US Freedom of Information Act (FOIA), then in its third decade, had been a catalyst for better government by exposing abuse, mismanagement, waste and violations of the law. It has actually saved money. Evans quoted the conclusions of Congressman Glenn English, who chaired a Congressional Inquiry into the effects of the act: "The savings that result from the FOIA disclosures are more than the costs of the Act. When the intangible benefits such as confidence, waste deterrence and a better informed citizenry are considered, the FOIA is a tremendous bargain. The FOIA more than pays for itself."[5] Neither Conservative nor Labour governments have ever explained why the British people cannot be trusted with such an act.

Margaret Thatcher has hailed the World Commission report as a "major and historic document". One of its main conclusions was that many governments need to recognize and extend citizens' and their organizations' rights "to know and have access to information on the environment and natural resources". The Thatcher government has neither recognized nor extended those rights. As long as Britain's Official Secrets Act, supported by a culture of government secrecy, remains in place, the British people will remain unable to participate fully in the national and global decisions now being made in their name.

There are other barriers to information besides the legal. In 1987, a Roper poll revealed that for the first time ever, half the US public cited television as their main source of news. Only 36 per cent cited newspapers.[6] A 1989 Harris poll found that 84 per cent of the British people gave television as their main source of information about developing countries and their problems. British television was the first to bring the realities of the Ethiopian famine into homes around the world in 1984.

Half-hour news programmes carry only a fraction of the information, in terms of the number of words, that appear on one page of a broadsheet newspaper. But television can bring issues alive in ways that, for a growing number of people, the printed word cannot.

British television has long been regulated to assure that serious programmes on serious issues get prime-time coverage. In the United States, television has been largely deregulated; the average home with cable television has 35–40 channels from which to choose. The average US viewer changes channels every three minutes. Programme ratings determine which programmes survive, so producers aim for the largest appeal.

Britain is moving towards less regulation, as seen in the government's broadcasting bill introduced in late 1989. It is complex and its expected results controversial. But it offers less regulation of quality and the timing of programmes, the inclusion of unregulated cable networks, and more emphasis on competition for viewers and for the sponsorship of programmes. Almost all major British charities concerned with the environment and the Third World – ranging from Oxfam to the United Nations Association to the WorldWide Fund for Nature (WWF) – have launched a campaign to change this law, to guarantee serious, prime-time information programmes.

"Public understanding of the dangers facing our planet and of the continuing poverty of millions has been greatly enhanced by broadcast coverage", said James Firebrace, editor of a charities-backed report on the issue. "There is now a real danger that this will be lost."

The leader in the first issue of the *British Journalism Review* agreed that television quality seemed set for decline, but added:

> What is perhaps less well perceived is the benign consequence for the government of television companies which are deprived of the interest and confidence and sheer sense of their own permanence which has, on occasion, made their journalism a mighty scourge. A government which auctions ITV off to the highest bidders, and prepares for the abolition of the BBC as we know it, is doing itself, in Machiavellian terms, a "useful political favour".[7]

It is hard to be for democracy and be against more freedom of choice on television. But improved information technology cannot be allowed to *deprive* people of the facts needed for informed choice.

THE POLITICS OF SUSTAINABLE DEVELOPMENT

The vision of sustainable development has the power to unite many different pressure groups under one common banner. The idea forever unites environmentalists and those concerned with progress in the South by showing that there can be no progress in a ravaged environment, and that progress based on short-sighted goals can only ravage the environment.

Already Oxfam, a development group, is campaigning on environmental issues because it realizes that villages in Africa cannot improve their economic lot in a bankrupt environment. The WorldWide Fund for Nature (WWF), traditionally a species protection organization, has widened its brief to worry about setbacks in development such as the debt crisis, because heavily indebted nations are forced to destroy forests and other natural habitats of species to make money.

Sustainable development is a "women's rights" issue, not only because women are the direct care-takers of the environment, as they do so much of the farming, wood collecting and water hauling in the Third World, but because "effective citizen participation in decision making" requires the effective participation of women.

Effective participation would also require the involvement of minority groups such as tribal peoples, especially those living in the rainforests and other fragile ecosystems. So sustainable development becomes an issue for groups fighting for the rights of these people.

Sustainable development includes peace groups, because it demands a new definition of "security" based on ecological security, which cannot be achieved with weapons. It also embraces freedom of information campaigners, who have always been concerned with access to environmental information.

As sustainable world progress will require levels of international co-operation never witnessed before on the planet, it is an issue which groups concerned with promoting such teamwork – such as the United Nations Associations – are also beginning to support.

This uniting of concern is not just a theoretical ideal. It is happening. Seven groups which had never before co-operated have formed a working group which has produced two reports on ways in which the British government ought to be promoting sustainable development. These included Friends of the Earth, Oxfam, Quaker Peace and Service, Survival International (tribal peoples' rights), United Nations Association-UK, World Development Movement, and WWF-UK.[8]

The *Global Tomorrow Coalition* in the United States represents a much larger collection of groups. In late 1989 it held the first-ever national public hearings on sustainable development, drawing over 1,000 delegates and producing a "US Citizens' Response to Sustainable Development" to send to Congress and to US and international leaders. In Geneva, the Centre for Our Common Future has drawn together 137 "working partners" worldwide, representing some of the most powerful citizens, industrial, trade union and scientific organizations.

WIDENING THE CIRCLE

But where is everyone else?

Global warming, ozone loss and the expected effects upon ourselves, our children and our grandchildren should make environmentalists of us all.

In Britain, some groups with the most obvious need to get involved in these issues have ignored them. For instance, the National Trust is Britain's biggest group concerned with the state of nation's environment, both built and natural; it has 1.7 million members. It has recently, in "Operation Neptune", purchased 500 miles of British coastline. Yet, amazingly, the National Trust has developed no policy on sea-level rise, sent no expressions of concern to the government, nor provided its members with any information on this threat. An embarrassed official could find nothing in the Trust's files on global warming or sea-level rise.

Once the well-being of children is added to the notion of sustainable development, as it must be, then the circle of concern should become infinitely wider and its members infinitely more active. It would include all parents, and everyone else concerned for children.

There is a move in this direction. The green consumerism of the late 1980s has done much to raise awareness of environmental issues in business, commerce and many groups and individuals who were not traditionally part of the environmental movement. Women's and parents' magazines, at all levels of the market, now carry regular green articles. Some of these go beyond the

Children facing the future boldly in Niger, West Africa. *Mark Edwards/ Still Pictures*

advertiser-led "green kitchen" and "green babies" features. The tabloid press has at last jumped on the green bandwagon.

It is too early in the history of green consumerism to tell whether this is part of the solution or part of the problem. If it stops short, having got a kick out of having switched to biodegradable nappies and chlorofluorocarbon (CFC)-free deodorants, then it will have done more harm than good, having deluded itself that it has changed things. But this may be only the beginning. People may be ready to go on and study products "from cradle to grave". Is their manufacture environmentally friendly? Do the companies that produce them increase poverty in other countries? Is a medicine or food sold in the North made by a company which pushes dangerous anti-diarrhoea drugs or encourages unnecessary bottle-feeding of babies in the Third World?

BRINGING THE WORLD TOGETHER

A government's basic job is to provide a system in which people can meet their own and their children's basic needs. Environmental destruction can wreck such a system, bringing with it poverty and ill health, especially for the poorest and most vulnerable families. It makes impossible the attainment of any real "national security", or any real "social security".

While there is little sign that governments are beginning to reorganize around this truth, to establish systems in which direct responsibility for the environment and the future permeates all law-making bodies and all ministries, there is some evidence that ordinary people and their organizations are.

Des Wilson, Britain's best known pressure group campaigner, argues that "nearly all of the great social changes in [Britain] were in fact advocated and fought for by pressure groups against the policies of the government of the day – from the abolition of slavery to votes for women, from the relief of poverty to greater civil rights".[9]

The anthropologist Margaret Mead wrote: "Never doubt that a small group of thoughtful, committed citizens can change the world. Indeed, it's the only thing that ever has."

Each of us is ultimately responsible for our government's policies. Either we know about and understand those policies and their effects or we do not. Once we understand, we either approve of those policies or we do not. And once we have made our minds up, we either communicate our views to the government and our friends or we do not.

Each of us is responsible for the ongoing debt crisis, for our nations' short-sighted energy use policies, for trade policies and for all other policies affecting children and the environment. The most highly educated and richest people on earth are the most wasteful consumers of resources and the worst polluters of the atmosphere. In this our governments back us.

Outrage is a powerful weapon for political change, and outrage is easy where children's suffering is involved. But outrage is usually short-lived. Love is more powerful and more lasting. Yet, try as we might, it is hard to love the planet and future generations in the abstract. Love of children comes more naturally. Children, as they become the focal point of the environment and development debate, can strengthen our resolve to make the difficult political choices necessary to sustain human life on earth.

REFERENCES

CHAPTER 1

1. World Commission On Environment and Development, *Our Common Future* (Oxford: Oxford University Press, 1987).
2. Clark, William, "Managing planet earth", *Scientific American*, vol. 261, no. 3, 1989.
3. McKibben, Bill, *The End of Nature* (London: Viking, 1990).

CHAPTER 2

1. Amosu, A., and J. Ozanne, "Band Aid – do they know it's finished?", *Financial Times*, 13 October 1989.
2. World Bank, *Sub-Saharan Africa: from Crisis to Sustainable Growth* (Washington, DC: World Bank, 1989).
3. UNICEF, *State of the World's Children 1989*, (Oxford: Oxford University Press, 1989).
4. Van der Hoeven, R., "Unwanted heritage: the threat of continuous debt overhang for children in the 21st century", UNICEF (unpublished), December 1989.
5. Bell, Emily, "One man's Third World crusade", *The Observer*, 4 February 1990.
6. MacPherson, Stewart, *Five Hundred Million Children, Poverty and Child Welfare in the Third World* (Brighton: Wheatsheaf Books, 1987).
7. George, Susan, *A Fate Worse than Debt* (London: Penguin, 1988).
8. Organization for Economic Co-operation and Development, *External Debt Statistics* (Paris: OECD, 1989).
9. UNICEF, *State of the World's Children 1990* (Oxford: Oxford University Press, 1990).
10. See note 9.
11. See note 9.
12. See note 9.
13. World Commission on Environment and Development, *Our Common Future* (Oxford: Oxford University Press, 1987).
14. See note 9.
15. Save The Children/Overseas Development Institute, *Prospects for Africa* (London: Hodder & Stoughton, 1988).
16. See note 2.
17. Cornia, A.C., R. Jolly, and F. Stewart (eds), *Adjustment with a Human Face*, vol. 2 (Oxford: Clarendon Press, 1988).
18. See note 3.
19. See note 9.
20. See note 9.
21. See note 9.
22. See note 15.
23. See notes 9 and 18.
24. See note 17.
25. Boyden, Jo, and Andy Hudson, *Children: Rights and Responsibilities*, Minority Rights Group, November 1985.
26. Vittachi, Anuradha, *Stolen Childhood* (Cambridge: Polity Press, 1989).
27. See note 25.

28. See note 26.
29. See notes 6 and 25.
30. Anti-Slavery Society, *Children in Especially Difficult Circumstances: The Sexual Exploitations of Children*, 1985.
31. See note 25.
32. Morch, J., "Abandoned street children", *Ideas Forum* no. 18, cit. Stewart MacPherson, *Five Hundred Million Children*, (Brighton: Wheatsheaf Books, 1987).
33. See note 25.
34. See note 25.
35. Amnesty International, *Iraq, Children: Innocent Victims of Political Repression*, February 1989.
36. Sivard, Ruth, *World Military and Social Expenditures 1989* (Washington, DC: World Priorities Inc., 1989).
37. Church of England, *Faith in the City: a Call For Action by Church and Nation* (London: Church House Publishing, 1985).
38. Coe, Richard, "US gets worst rating of 12 industrial states for welfare of children", *International Herald Tribune*, 20 March 1990.
39. Mintel, *British Lifestyles 1990* (London: Mintel Publications, 1990).
40. House of Commons, Hansard, 29 July 1988, cols 820–21.
41. Child Poverty Action Group, "Poverty: the facts" (London: CPAG Ltd, 1988).
42. Burghes, L., *Children in Poverty* (London: National Children's Bureau, 1984).
43. Whitehead, M., *The Health Divide: inequalities in health in the 1980s* (London: Health Education Council, 1988).
44. Low Pay Unit, *Ten Years On: the Poor Decade* (London: Low Pay Unit, 1989).
45. *Mortality Statistics 1985*, HMSO, 1988 cit. Child Poverty Action Group, *Poverty the Facts* (London: CPAG Ltd, 1988).
46. "Black Americans", *The Economist*, 3 March 1990.
47. The Black Report, *Inequalities in Health*, Department of Health and Social Security, 1980.
48. Whitehead, Margaret, *The Health Divide: Inequalities in Health in the 1980s* (London: Health Education Council, 1987).
49. See note 48.
50. Pollitt, Nigel, "Hard times, young and homeless", Shelter, 1989.
51. *Social Trends 20* (London: HMSO, 1990).
52. Cole-Hamilton, Issy, "Homeless and hungry", *The Food Magazine*, London Food Commission, summer 1988.
53. Audit Commission, Shelter, London Housing Unit cit. Michael Tintner, *State Imperfect: the book of social problems* (London: Macdonald Optima, 1989).
54. See note 51.
55. Confederation of British Industry, *Towards a Skills Revolution: Report of the Vocational Education and Training Task Force*, November 1989.
56. *Standards in Education, 1988–89: the Annual Report of the HM Senior Chief Inspector of Schools*, Department of Education & Science, January 1990.
57. Lloyds Bank Economic Bulletin cit. Peter Jenkins, "At the bottom of the class", *The Independent*, 8 February 1990.
58. See note 56.
59. Jenkins, Peter, "At the bottom of the class", *The Independent*, 8 February 1990.
60. See note 51.
61. *Regional Trends 23* (London: HMSO, 1988).
62. See note 51.

63. Cohen, Bronwen, *Caring for Children*, Commission of the European Communities, 1988.
64. Kempe, C.H. *et al.*, "The Battered Child Syndrome", *Journal of the American Medical Association*, no. 181, 1962.
65. Tintner, Michael, *State Imperfect, the book of social problems* (London: Macdonald Optima, 1989).
66. See note 51.
67. Le Vine, S. and R. Le Vine, "Child abuse and neglect in sub-Saharan Africa" in J. Korbin (ed), *Child Abuse and Neglect: Cross-Cultural Perspectives* (Berkeley: University of California Press, 1981).
68. Korbin, J., *Child Abuse and Neglect: Cross-Cultural Perspectives* (Berkeley: University of California Press, 1981).
69. *The Food Magazine*, The Food Commission, April 1990.
70. World Health Organization, *World No-Tobacco Day, Childhood and Youth without Tobacco*, Geneva, May 1990.
71. Lapham, Lewis, "A political opiate", *Harper's Magazine*, December 1989.
72. "US says health cost of smoking cigarettes is $52 billion a year", *International Herald Tribune*, 22 February 1990.
73. Stephen, Andrew, "Haunted by the grim reaper of Winston-Salem", *The Observer*, 25 February 1990.
74. See note 70.
75. United Nations Environment Programme, *UNEP Environmental Data Report 1989–90* (Oxford: Blackwell, 1989).
76. See note 75.
77. Clark, Eric, *The Want Makers* (London: Hodder & Stoughton, 1988).
78. Wilson, Des, *Parents Against Tobacco Handbook* (London: Parents Against Tobacco, 1990).
79. See note 78.
80. See note 77.
81. See note 77.

CHAPTER 3

1. Timberlake, Lloyd, "Poland – the most polluted country in the world?", *New Scientist*, 22 October 1981.
2. "Clearing up after communism", *The Economist*, 17 February 1990.
3. "Poison politics", *Newsweek*, 22 January 1990.
4. Wohl, Anthony S., *Endangered Lives, Public Health in Victorian Britain* (London: J.M. Dent & Sons Ltd, 1983).
5. Kurzel, R.B. and C.L. Cetrulo, "The effect of environmental pollutants on human reproduction, including birth defects", *Environmental Science and Technology*, vol. 15, 1981.
6. "Disabilities in US babies found to have doubled in 25 years", *International Herald Tribune*, 31 July 1983, cit. John Elkington, *The Poisoned Womb* (London: Viking, 1985).
7. Elkington, John, *The Poisoned Womb* (London: Viking, 1985).
8. Rogan, W.J. *et al.*, "Congenital poisoning by polychlorinated biphenyls and their contaminants in Taiwan", *Science*, vol. 241, p. 334, 1988.
9. Skerfvig, S., "Mercury in women exposed to methylmercury through fish

consumption, and in their newborn babies and breast milk", *Bulletin of Environmental Contamination Toxicology*, vol. 41, 1988.
10. Timberlake, Lloyd, *Only One Earth* (London: BBC Books, 1987).
11. Needleman *et al.*, "Deficits in psychological and classroom performance of children with elevated dentine lead levels", *New England Journal of Medicine*, vol. 300, 1979.
12. *Octel News*, May 1981 (cit. Des Wilson, *The CLEAR Handbook, Campaign for Lead-free Air*, 1982).
13. Wilson, Des, *The Lead Scandal* (London: Heinemann, 1983).
14. See note 13.
15. Russell Jones, Dr Robin, "The continuing hazard of lead in drinking water", *The Lancet*, vol. 2, September 1989.
16. United Nations Environment Programme, *UNEP Environmental Data Report 1989–90* (Oxford: Blackwell, 1989).
17. See note 15.
18. Fulton, M., *et al.*, "Influence of blood lead on the ability and attainment of children in Edinburgh", *The Lancet*, vol. 1, May 1987.
19. "Drinking water regulations; maximum contaminant level goals & national primary drinking water regulations for lead and copper; proposed rule", Part V, Environmental Protection Agency, Federal Register, vol. 53, no. 160, 18 August 1988.
20. Food Surveillance Paper no. 27, *Lead in Food: Progress Report*, The 27th Report of the Steering Group on Food Surveillance, The Working Party on Inorganic Contaminants in Food, Third Supplementary Report on Lead, HMSO, London, 14 December 1989.
21. See note 15.
22. Friends of the Earth, *Earth Matters*, no. 6, Winter 1989/90.
23. Berwick, Ian, "Unleaded petrol advances through Europe", *Petroleum Review*, July 1989.
24. Global Environment Monitoring Service, *Assessment of Urban Air Quality* (Geneva: UNEP/WHO, 1988).
25. French, Hilary, "Clearing the air", *State of the World 1990* (New York: Norton, 1990).
26. Breach, Ian, "Can you give up your car?", *The Guardian*, 13 October 1989.
27. "Prime Minister has abandoned green concerns", *The Independent*, 17 March 1990.
28. "Cars – the facts", *New Internationalist*, May 1989.
29. See note 24.
30. Transport 2000, *Vital Travel Statistics 1990* (London, 1990).
31. "More money for jam", *The Times*, 1 December 1989.
32. *Social Trends 20* (London: HMSO, 1990).
33. Earth Resources Research *et al.*, "Company cars: a pre-budget briefing", *Earth Resources Research*, 1990.
34. Transport 2000, *Transport Retort*, Jan/Feb 1990.
35. See note 32.
36. "Company cars: the hidden cost to Britain", *The Observer*, 7 January 1990.
37. See note 34.
38. See note 36.
39. Friends of the Earth, "Getting there", National Transport Policy, Friends of the Earth, 1987.
40. See note 32.

41. "The price that motorists must pay", *The Independent*, 5 October 1989.
42. Lowe, Marcia, "Cycling into the future", *State of the World 1990* (New York: Norton, 1990).
43. "Roads for prosperity", Department of Transport, London, May 1989.
44. Friends of the Earth, *Kids Alive, Accident Facts and Figures* (London, 1987).
45. See note 32.
46. World Commission in Environment and Development, *Our Common Future* (Oxford: Oxford University Press, 1987).
47. Patten, Chris, Speech at Policy Studies Institute, 15 February 1990.
48. United Nations Environment Programme (UNEP), *Environmental Data Report 1989–90* (Oxford: Blackwell, 1989).
49. See note 25.
50. Holman, Claire, *Air Pollution and Health* (London: Friends of the Earth, 1989).
51. Children's Legal Centre, *Children and the Environment*, London, 1989.
52. See note 50.
53. See note 50.
54. *Digest of Environmental Protection and Water Statistics*, no. 11, 1988 (HMSO, 1989).
55. See note 24.
56. See note 24.
57. See note 54.
58. See note 24.
59. Holman, Claire, *Particulate Pollution from Diesel Vehicles* (London: Friends of the Earth, 1989).
60. See note 24.
61. See note 24.
62. See note 25.
63. World Resources Institute/International Institute on Environment and Development, *World Resources 1988–89* (New York: Basic Books, 1988).
64. See note 50.
65. See note 24.
66. See note 24.
67. See note 24.
68. See note 25.
69. "Clearing up after communism – East European pollution", *The Economist*, 17 February 1990.
70. See note 24.
71. de Koning, H., "Air pollution in Africa", *World Health*, Jan/Feb 1990.
72. See note 51.
73. "Control of pollution (special waste) regulations 1980" cit. *Children and the Environment* (London: Children's Legal Centre, 1989).
74. "Ministers accused of lax attitudes to contaminated land", *New Scientist*, 10 February 1990.
75. Department of the Environment, "Assessment of groundwater quality in England and Wales" (London: HMSO, 1988) cit. Friends of the Earth Briefing Sheet, *Sitting on a Pollution Time Bomb* (London: Friends of the Earth, 1990).
76. Friends of the Earth, *The Government's Environmental Record*, London, December 1989.
77. Erlichman, James, "Babies 'at risk from nitrates in water'", *The Guardian*, 3 October 1988.
78. See note 72.
79. See note 76.

80. Friends of the Earth Briefing Sheet, *Nitrates*, London, 1988.
81. National Academy of Sciences, *Regulating Pesticides in Food – the Delaney Paradox* (Washington, DC: National Academy Press, 1987).
82. Sewell, B., R. Whyatt, J. Hathaway, and L. Mott, *Intolerable Risk: Pesticides in Our Children's Food* (New York: National Resources Defense Council, 1989).
83. Erlichman, James, "How Alar was praised and then condemned", *The Guardian*, 20 October 1989.
84. See note 82.
85. Friends of the Earth/Parents for Safe Food press release, "Dangerous agrochemical in supermarket foods", 14 February 1990.
86. Advisory Committee on Pesticides, "Position document on consumer risk arising from the use of EBDCs", Ministry of Agriculture, January 1990.
87. See note 82.
88. Albert, Lilia, "Children and pesticides in Mexico", *Journal of Pesticide Reform*, vol. 9, no. 3, 1989.
89. Erlichman, James, "Pesticide tested on children", *The Guardian*, 12 December 1982; James Erlichman, "How the drug giants profit when life is cheap", *The Guardian*, 19 January 1983.
90. See note 82.
91. Garland, Anne Witte, *For Our Kids' Sake: How to Protect Your Child Against Pesticides in Food* (New York: Natural Resources Defense Council, 1989).
92. *The Independent*, 5 March 1990.
93. Environmental Data Services Ltd, ENDS Report 170, March 1989.
94. See note 82.
95. Department of the Environment, "Dioxins in the environment", Pollution Paper no. 27, HMSO, 1989.
96. Caulfield, Catherine, *Multiple Exposures* (London: Secker & Warburg Ltd, 1989).
97. Ingrams, Richard, *The Observer*, 18 February 1990.
98. Gardner, Martin *et al.*, "Results of case-control study of leukaemia and lymphoma among young people near Sellafield nuclear plant in West Cumbria", *British Medical Journal*, vol. 300, 17 February 1990.
99. "Danger: men at work", *The Economist*, 10 March 1990.
100. Morris, Michael, "Cumbrians calm on Sellafield report", *The Guardian*, 17 February 1990.
101. National Radiation Protection Board, *Interim Guidance on the Implications of Recent Revisions of Risk Estimates and the ICRP 1987 Como Statement* (London: HMSO, 1987).
102. "Radiation risks underestimated", Friends of the Earth press release, 30 November 1988.
103. Milne, Roger, "Leukaemia study sparks review of radiation limits", *New Scientist*, 24 February 1990.
104. Sowby, F.D., "ICRP dose limits from 1934 up to 1977", *Radiological Protection Bulletin*, no. 28, May 1979, cit. Catherine Caulfield, *Multiple Exposures* (London: Secker & Warburg Ltd, 1989).
105. ICRP–26 recommendations, 1977, cit. Catherine Caulfield, *Multiple Exposures* (London: Secker & Warburg Ltd, 1989).
106. Thorne, Michael, ICRP's then scientific secretary in an interview with Catherine Caulfield, 12 September 1985, cit. Catherine Caulfield, *Multiple Exposures* (London: Secker & Warburg Ltd, 1989).
107. "Indian nuclear plants accused over radioactive discharges", *New Scientist*, 17 February 1990.
108. See note 103.

109. Baxter, M.S., *et al.*, "A review of radioactivity in and around the Capper Pass Smelter, Melton Works, North Humberside", Scottish Universities Research and Reactor Centre, Glasgow, March 1990.
110. See note 109.
111. See note 96.
112. National Radiological Protection Board, "Living with radiation", NRPB, 1989.
113. See note 99.
114. See note 51.

CHAPTER 4

1. World Health Organization, *World Health*,May 1988.
2. UNICEF, *The State of the World's Children 1990* (Oxford: Oxford University Press, 1990).
3. Read, Kathy, "Behind the face of malnutrition", *New Scientist*, 17 February 1990.
4. World Health Organization, *Global Nutritional Status* (Geneva: WHO, 1989).
5. See note 4.
6. Baile, Stephanie, "Women and health in developing countries", *The OECD Observer*, Dec 1989/Jan 1990.
7. Morley, D., and H. Lovel, *My Name is Today* (London: Macmillan, 1986).
8. World Health Organization, "Infant and young child nutrition", executive board paper EB85/18, 8 December 1989.
9. Palmer, Gabrielle, *The Politics of Breastfeeding* (London: Pandora Press, 1988).
10. "Breastfeeding as a family planning model", *Mothers and Children*, vol. 8, no. 1, 1989.
11. See note 2.
12. See note 7.
13. See note 2.
14. World Health Organisation/UNICEF, *International Code of Marketing of Breast-milk Substitutes* (Geneva: WHO, 1981).
15. "Nestlé: the boycott's back", *Multinational Monitor*, September 1988.
16. Bond, Gwenda, "Where there can be pain in a packet of milk", *Baptist Times*, 26 October 1989.
17. Baby Milk Action Coalition, *BMAC Newsletter*, Summer 1989.
18. World Health Assembly Resolution WHA39.28, May 1986.
19. See note 8.
20. King, Thomas, "Doctors vow to proscribe infant-formula ad plans", *Wall Street Journal*, 24 August 1989.
21. See note 2.
22. UNICEF, World Health Organization and UNESCO, *Facts for Life* (New York: UNICEF, 1989).
23. Wohl, Anthony, *Endangered Lives: Public Health in Victorian Britain* (London: J.M. Dent & Sons, 1983).
24. "Iodine", *World Health*, May 1988.
25. UNICEF, *The State of the World's Children 1989* (Oxford: Oxford University Press, 1989).
26. Wingate, Paul, *Penguin Medical Dictionary* (2nd edn), (London: Penguin, 1976).
27. See note 2.
28. Christie, Debbie (producer), "Hard to swallow", *World in Action*, 17 October 1989.
29. Health Action International, *Problem Drugs* (The Hague: HAI, 1986).
30. "Notes and news" *The Lancet*, 25 November 1989.
31. See note 28.
32. Walsh, J.A., and K.S. Warren, *New England Journal of Medicine*, 1979 vol. 301, p. 967,

cit. Agarwal *et al.*, *Water Sanitation and Health – for All?* (London: Earthscan, 1981).
33. Cairncross, Sandy, "Water!", *World Health*, January/February 1990.
34. See note 2.
35. United Nations Environment Programme, *Sands of Change*, Environment Brief no. 2 (Nairobi: UNEP, 1987).
36. Organization for Economic Co-operation and Development, *The Sahel Facing the Future* (Paris: OECD, 1988).
37. UNICEF, *Within Human Reach: a Strategy for Africa's Children* (New York: UNICEF, 1985).
38. Jayal, Nalini, "Destruction of water resources – the most critical ecological crisis of East Asia", paper presented to IUCN conference, Madrid 1984.
39. Giri, Jacques, "Retrospective de l'économie Sahélienne" (Paris: Club du Sahel, 1984).
40. See note 36.
41. Ellis, William, "A Soviet sea lies dying", *National Geographic*, vol. 177, no. 2, February 1990.
42. Cornwell, Rupert, "Catastrophe on Aral Sea 'ranks with Chernobyl'", *The Independent*, 20 April 1989.
43. "Wasteland: the Soviet environmental nightmare", *Newsweek*, 22 January 1990.
44. See note 43.
45. Poore, Duncan, *No Timber Without Trees* (London: Earthscan Publications Ltd, 1989).
46. United Nations Environment Programme, *Environmental Data Report 1989–90* (London: Blackwell, 1989).
47. World Resources Institute/International Institute for Environment and Development, *World Resources 1988–89* (New York: Basic Books, 1988).
48. Kumar, S., and D. Hotchkiss, *Consequences of Deforestation for Women's Time Allocation, Agricultural Production and Nutrition in Hill Areas of Nepal* (Washington, DC: International Food Policy Research Institute, 1988).
49. See note 46.
50. See note 47.
51. World Commission on Environment and Development, *Our Common Future* (Oxford: Oxford University Press, 1987).
52. Myers, Norman (ed), *The Gaia Atlas of Planet Management* (London: Pan Books,1985).
53. "The decade by numbers", *Harper's Magazine*, January 1990.
54. See note 51.
55. HRH Prince Charles, The Rainforest Lecture, Royal Botanic Gardens, Kew, 6 February 1990.
56. Colchester, Marcus, "Unaccountable aid: secrecy in the World Bank", *Index on Censorship*, vol. 18, nos 6 & 7, July/August 1989.
57. See note 56.
58. Hardoy, J., and D. Satterthwaite, *Squatter Citizen* (London: Earthscan Publications Ltd, 1989).
59. See note 51.
60. Wohl, Anthony, *Endangered Lives: Public Health in Victorian Britain* (London: J.M. Dent & Sons, 1983).
61. World Health Organization, *Urbanization and its Implications for Child Health* (Geneva: WHO, 1988).
62. Rodhe, J.E., "Why the other half dies: the science and politics of child mortality in the Third World", *Assignment Children*, vol. 61–2, 1983.
63. See note 61.

64. McAuslan, Patrick, *Urban Land and Shelter for the Poor* (London: Earthscan, 1985).
65. See note 58.

CHAPTER 5

1. Duhl, Len, "Health and the city", *World Health*, January/February 1990.
2. Thomas, Hugh, *An Unfinished History of the World* (3rd edn), (London: Pan Books, 1989).
3. Cooke, Henrietta, "Third World winning measles war", *The Observer*, 18 February 1990.
4. UNICEF, *The State of the World's Children 1989* (Oxford: Oxford University Press, 1989).
5. UNICEF, "Strategies for children in the 1990s", executive board paper E/ICEF/1989/L.5, 1989.
6. Bland, John, ed., *World Health*, World Health Organization, December 1989.
7. Henderson, Donald, "How smallpox showed the way", *World Health*, December 1989 (Geneva, WHO).
8. John, T. Jacob, "Where the need is greatest", *World Health*, December 1989 (Geneva, WHO).
9. Goldsmith, G., and N. Hildyard (eds), *The Earth Report* (London: Mitchell Beazley, 1988).
10. Levy, S.B., "Playing antibiotic pool; time to tally the score", *New England Journal of Medicine* vol. 311, no. 10, 1984.
11. "Overdosing on antibiotics", *Newsweek*, 17 August 1981.
12. *ARI News*, April 1989, Appropriate Health Resources and Technology Action Group, London.
13. Douglas, Robert, *ARI News*, August 1986, Appropriate Health Resources and Technology Action Group, London.
14. See note 4.
15. See note 4.
16. Vitousek, Peter, *et al.*, "Human appropriation of the products of photosynthesis", *Bioscience*, vol. 34, no. 6, May 1986.
17. Daly, Herman, "Sustainable development: from concept and theory towards operational principles", Hoover Institute Conference on Population and Development Review, March 1988.
18. Wahren, Carl, "Population and development – the burgeoning billions", *The OECD Observer*, December 1988/January 1989.
19. Gille, Halvor, "The world fertility survey: policy implications for developing countries", *International Family Planning Perspectives*, vol. 11, no. 1, March 1985.
20. See note 4.
21. Lappé, F., and R. Schurman, *Taking Population Seriously* (London: Earthscan Publications Ltd, 1989).
22. See note 21.
23. Earthscan, *Primary Health Care*, Earthscan press briefing document no. 9 (London: Earthscan, 1978).
24. Mahler, H., and H. Labouisse, *Primary Health Care* (Geneva: World Health Organization, 1978).
25. Walsh, J.A., and K.S. Warren, "Selective primary health care: an interim strategy for disease control in developing countries", *New England Journal of Medicine*, vol. 301, no. 18, 1979.
26. Wisner, Ben, "GOBI versus PHC: some dangers of selective primary health care",

Social Science and Medicine, vol. 26, no. 9, 1988.
27. Rifkin, Susan, and Gill Walt, "Why health improves: defining the issues concerning 'comprehensive primary health care' and 'selective primary health care'", *Social Science and Medicine*, vol. 23, no. 6, 1986.
28. Kabell, Dorte, "What progress on primary health care?", *The OECD Observer*, December 1989/January 1990.
29. Taylor, Carl, and Richard Jolly, "The straw men of primary health care", *Social Science and Medicine*, vol. 26 no. 9, 1988.
30. See note 28.

CHAPTER 6

1. United Nations Environment Programme, *Environmental Data Report 1989–90* (Oxford: Basil Blackwell Ltd, 1989).
2. United Nations Environment Programme, *Environmental Effects of Ozone Depletion* (Nairobi: UNEP, 1989).
3. See note 1.
4. Parry, Martin, *et al.*,"Draft report on agriculture, forestry and land use", IPCC Working Group II, unpublished.
5. World Bank, *Poverty and Hunger* (Washington, DC: World Bank, 1986).
6. See note 2.
7. World Resources Institute/International Institute for Environment and Development, *World Resources 1988–89* (New York: Basic Books Inc., 1988).
8. See note 2.
9. McKibben, Bill, *The End of Nature* (London: Viking, 1990).
10. Weihe, W., *Climate, Health and Disease* (Geneva, World Meteorological Organisation, 1979).
11. World Meteorological Organisation, *Climate and Human Health* (Geneva: WMO, 1987).
12. Agarwal, Anil, *Mud, Mud* (London: Earthscan, 1981).
13. See note 2.
14. Grant, Lester, *Health Effects Issues Associated with Regional and Global Air Pollution Problems*, proceedings of Conference on The Changing Atmosphere: Implications for Global Security (Geneva: World Meteorological Organization, 1989).
15. See note 7.
16. Penkett, Stuart, "Ultraviolet levels down not up", *Nature*, vol. 341, 28 September 1989.
17. Expert Group, *Climate Change: Meeting the Challenge* (London: Commonwealth Secretariat, 1989).
18. Kalkstein, S., *et al.*, "The impact of human induced climatic change upon human mortality: a New York case study", *Effects of Changes in Stratospheric Ozone and Global Climate*, vol. 4, UNEP/EPA 1986.
19. ICF/Clement Associates, "The potential impact of climate change on patterns of infectious disease in the USA" (Washington, DC: US Environmental Protection Agency, 1987).
20. See note 1.
21. World Health Organization, *World Health*, December 1989.
22. See note 14.
23. Channing, Rory, "Global warming likely to produce new boat people", *Los Angeles Times*, 30 July 1989.

24. Leaf, Alexander, "Potential health effects of global climatic and environmental changes", *The New England Journal of Medicine*, vol. 321, no. 23.
25. Gribbin, John, "Did the greenhouse effect cause the storm?", *New Scientist*, 3 February 1990.
26. Seaman, John, *Epidemiology of Natural Disasters* (London: Karger, 1984).
27. See note 26.
28. O'Neil, Bill, "Cities against seas", *New Scientist*, 3 February 1990.
29. Kates, R.W., *et al.*, "Human impact of the Managua earthquake", *Science*, vol. 182, 1973.
30. Wijkman, Anders, and Lloyd Timberlake, *Natural Disasters: Acts of God or Acts of Man?* (London: Earthscan, 1984).
31. Sommer, A., and W.H. Mosely, "East Bengal cyclone of November 1970 – epidemiological approach to disaster assessment", *Lancet* pp. 1029–36, 1978.
32. Glass, R.I., *et al.*, "Earthquake injuries related to housing in a Guatemalan village", *Science*, vol. 197, pp. 638–43, 1977.
33. See note 26.
34. Milliman, John, "Environmental and economic implications of rising sea level and subsiding deltas: the Nile and Bengal Examples", *Ambio*, vol. 18, no. 6, 1989.
35. Timberlake, Lloyd, "Global weather: what on earth is happening?", *Telegraph Weekend Magazine*, 5 November 1988.
36. Sinclair, Jan, "Rising sea levels could affect 300 million", *New Scientist*, 20 January 1990.
37. World Commission on Environment and Development, *Our Common Future* (Oxford: Oxford University Press, 1987).
38. See note 17.
39. "The decade by numbers", *Harper's Magazine*, January 1990.
40. Lean, Geoffrey, "Revealed: 1,350 poison dumps threaten water", *Observer*, 4 February 1990.
41. See note 1.
42. See note 17.
43. Jaeger, Jill, "Developing policies for responding to climatic change", prepared for World Climate Programme, February 1988.
44. Gleick, Peter, "Climate change and international politics: problems facing developing countries", *Ambio*, vol. 18, no. 6, 1989.
45. Miller, Jack, "Chinese bring a chill to backers of ozone protocol", *New Scientist*, 11 February 1989.
46. "A worldwide problem", *The International Herald Tribune*, 12 February 1990.
47. Barbier, Edward, and David Pearce, "Thinking economically about climate change", *Energy Policy*, January/February 1990.
48. McKinsey and Co., *Protecting the Global Atmosphere: Funding Mechanisms*, Second Interim Report to Steering Committee for Ministerial Conference on Atmospheric Pollution and Climate Change, The Netherlands, 6–7 November 1989.
49. Brown, Lester, *State of the World 1988* (New York: Norton, 1988).
50. Roberts, Simon, "Setting standards for energy efficiency", Friends of the Earth, 1989.
51. Pearce, Fred, "Britain isolated over response to greenhouse effect", *New Scientist*, 25 March 1989.
52. See note 37.
53. Keepin, Bill and Gregory Kats, *Memorandum of Evidence on Energy Policy Implications of the Greenhouse Effect for the House of Commons Energy Committee* (London: HMSO, July 1989).

54. See note 37.
55. Schneider, Stephen, "The changing climate", *Scientific American*, vol. 261, no. 3, September 1989.

CHAPTER 7

1. McKibben, Bill, *The End of Nature* (London: Viking, 1990).
2. Timberlake, Lloyd, *Africa in Crisis* (London: Earthscan Publications Ltd, 1988).
3. Timberlake, Lloyd, "Poland – the most polluted country in the world?", *New Scientist*, 22 October 1981.
4. Sivard, Ruth, *World Military and Social Expenditures, 1989* (Washington, DC: World Priorities Inc., 1989).
5. Brown, Lester, *et al.*, *State of the World 1989* (New York and London: W.W. Norton & Co., 1989).
6. Boyle, Stewart, "Energy efficiency", *A Report on the Prime Minister's First Green Year* (London: Media Natura, 1989).
7. World Bank, *World Development Report 1980* (Oxford: Oxford University Press, 1980).
8. World Commission on Environment and Development, *Our Common Future* (Oxford: Oxford University Press, 1987).
9. Pearce, David, *et al.*, *Blueprint for a Green Economy* (London: Earthscan Publications Ltd, 1990).
10. Freeman, M., *Air and Water Pollution Control: a Cost-Benefit Assessment* (New York: Wiley, 1982).
11. Maltby, Edward, *Waterlogged Wealth* (London: Earthscan Publications Ltd, 1986).
12. Anderson, Dennis, *The Economics of Afforestation* (Baltimore: Johns Hopkins University Press, 1987).
13. Weiss, E.B., *In Fairness to Future Generations* (Tokyo: United Nations University, 1989).
14. See note 13.
15. Pérez de Cuéllar, P., message to International Meeting on the UN Convention on the Rights of the Child, Lugano, Italy, 1987.
16. See note 8.
17. See note 8.
18. "The environment", *The Economist*, 2 September 1989.
19. See note 9.
20. UNICEF, *State of the World's Children 1990* (Oxford: Oxford University Press, 1990).

CHAPTER 8

1. *Gallup Poll News Service*, vol. 54, no. 3, 17 May 1989.
2. "Worldwide concern about the environment", *Our Planet*, UNEP, vol. 1, no. 2/3, 1989.
3. Montalbano, W., "Green wave surges over Europe", *Los Angeles Times*, 11 May 1989.
4. Ignatieff, Michael, "Tough lessons for true democrats", *The Observer*, 11 March 1990.
5. Evans, Harold, address to Freedom of Information Awards Ceremony, London, 30 January 1989.
6. *The Progressive Review*, no. 264, May 1987, Washington, DC.
7. "Why we are here", *British Journalism Review*, vol. 1, no. 1, Autumn 1989.

8. International Institute for Environment and Development, *Britain and the Brundtland Report* and *Brundtland in the Balance* (London: IIED, 1989).
9. Wilson, Des, "Address to the Howard League", 17 November 1988.